A Lakeside Companion

The University of Wisconsin Press

TED J. RULSEH

A

LAKESIDE

·COMPANION·

The University of Wisconsin Press
1930 Monroe Street, 3rd Floor
Madison, Wisconsin 53711-2059
uwpress.wisc.edu

3 Henrietta Street, Covent Garden
London WC2E 8LU, United Kingdom
eurospanbookstore.com

Printed in the United States of America

This book may be available in a digital edition.

Library of Congress Cataloging-in-Publication Data

Names: Rulseh, Ted, author.
Title: A lakeside companion / Ted J. Rulseh.
Description: Madison, Wisconsin: The University of Wisconsin Press, [2018]
| Includes bibliographical references and index.
Identifiers: LCCN 2018011136 | ISBN 9780299320003 (cloth: alk. paper)
Subjects: LCSH: Lakes. | Lake ecology.
Classification: LCC QH98 .R79 2018 | DDC 577.63—dc23
LC record available at https://lccn.loc.gov/2018011136

Earlier versions of this material appeared in columns published by the *Lakeland Times* and the *Northwood River News* and are published here by permission.

To my wife
NOELLE

but most of all to our children
SONYA and TODD
and our grandsons
TUCKER and PERRIN
for whose benefit our lakes and resources
must be preserved and protected.

Contents

Preface

When you look at your favorite lake from your window or pier, or from a campsite or resort, what do you see? Beautiful blue water? A place for a refreshing dip in summer's heat? A surface on which to paddle a canoe or kayak, float on an air mattress, or cruise in a boat? Favorite spots to catch fish for sport or the dinner table?

Your lake is all this, but it's also much more. It's a collection of worlds: in the water; in the sand, gravel, rocks, and muck of the bottom; on the surface; in the air above; and along the shoreline, a belt of land incredibly rich in life. All these intertwine. Just as important is what you can't see—the physical, biological, and chemical processes that determine, for example, how many and what kinds of fish live in the lake, which plants grow there and how profusely, the color and clarity of the water, and how soon the ice forms in winter and melts in spring.

There's much about lakes to know and understand, as you'll discover on these pages. I'm not a lake scientist—just someone who loves lakes, cares about them deeply, and has studied them in high school and college courses, in books and magazines and online sources, and through personal experiences. I grew up in Two Rivers, Wisconsin, on the shore of Lake Michigan, but I found inland lakes more captivating than the big lake, more accessible and inviting, especially those in the northern regions. The fascination took hold from the first time my family vacationed in a rustic lakefront cabin in the big woods of Upper Michigan, back when I was nine years old. I now live on Birch Lake, 180 acres of water, deepest point 27 feet, in the glacial lake country of north central Wisconsin.

Over the years I've fished large lakes from the comfort of a boat and smaller ones from a canoe or float tube, or in waders. I've snorkeled clear lakes, sliding like an otter over the smooth, barkless trunks of long-fallen pines to spy on bass and bluegills around sunken tangles of timber. I've watched sunsets over the water, observed loons and eagles, taken water-clarity readings as a volunteer lake monitor, paddled the shallows in a canoe at dawn. In short, I've spent considerable time exploring and learning about lakes and lake life.

No doubt you have enjoyed lakes in many of the same ways. Now I invite you to look deeper—at the forces that shape lakes and the life that abounds in them. Did you know, for example, that your lake's water has layers? That its water is really the broth of a thin soup rich in tiny plants and animals on which larger creatures feed? That the walleye or bass you catch owes its existence, first and foremost, to the sun? That lake ice melts from the bottom up? That lakes and the water under the ground are not separate entities but interconnected parts of the same system? That fish can breathe in water even though it holds less than one ten-thousandth as much oxygen as the air?

Here you'll learn about all this and much else in simple terms you don't have to be a scientist to understand. You'll also share, through my eyes, glimpses of life here on Birch Lake that may call to mind experiences you've had, or would like to have, on your favorite body of water. So come along. I hope this book helps you know your lake more intimately, and come to love and appreciate it even more than you do today.

A Lakeside Companion

Paul Skawinski, University of Wisconsin–Stevens Point

BASICS

·1·

PUTTING IN THE PIER

The winter ice is good for making sure we don't take our lakes for granted. The opening of the water is like a reunion with a beloved friend, and the official welcome is putting in the pier. Here on Birch Lake, the water usually opens in mid-April; sometimes the last of the ice lies up against our shore, postponing my pier project, but only for a day or so.

I look forward to this ritual much as I did putting up the Christmas tree when the kids were young, only this perhaps is better for all it portends. Maybe you roll your pier into the water on wheels. Perhaps you maneuver whole sections of light aluminum frame. Maybe your pier stays in the water all year. My pier goes together like pieces from an Erector set, and that adds to the childlike joy I feel in assembling it.

Preparations are easy. Sweatpants over trousers inside rubber hip boots—that water is cold! Hooded sweatshirt. Ball cap and sunglasses. Carpenter's square and level. Adjustable wrench. Ratchet wrench with socket. Plastic zipper-bag of bolts, nuts, and washers. And down the fifty-seven lakefront stairs I go.

It's in my nature to rush through chores like leaf raking and garage cleaning, but I don't hurry this one. I savor each step, working with care. The pier stands are stacked on shore in the order removed last fall. So, last out, first in. Set the first stand roughly 10 feet from the shore anchor. Balance it in place. Go get a side rail. Bolt one end to the shore anchor, the other to the stand, just finger tight. Repeat with opposite rail.

Check with the square; move the stand left or right as needed. Check with the level. Usually it's fine, since the same parts go in the same places as last year and the lake bottom of silty sand doesn't change much. Tighten down the nuts at the four corners, enjoying the staccato ring of the ratchet. Continue with the remaining three frame sections and then the L extension where the bench goes. While I work, sometimes an otter swims by and assumes a seat on a branch of the fallen pine tree along the shore of the neighboring lot. His growls tell me that for whatever reason he does not approve of my work.

Simple as this task may be, it never goes perfectly. A stand tips over with a splash. A side rail slips into the water. A nut escapes my grip, always at the deep end of the pier, and I have to fish it out, getting a soggy sleeve for the trouble. Then the frame is done. Laying on the sections of cedar decking is like building a plank road to a season of promise. Decking in place, bench installed, I sit for a while. The ice having left early, we can expect more than seven months of open water. It starts now, today. Imagine the adventures, the simple joys, the wonders that await.

HOW THE GLACIERS DID THEIR WORK

We know that glaciers formed most of the lakes where we live, carving out depressions in the earth, which then filled with water as the ice melted. But to understand more clearly how it happened, it helps to know how the glaciers moved, because without their slow and unstoppable movement, they wouldn't have carved anything.

It's easy to envision the movement of alpine glaciers, those that form in the mountains and are influenced by gravity. More mysterious is how the vast ice sheets moved across this landscape thousands of years ago. Those glaciers didn't move downhill. South may be represented by down on a map or a globe, but it's not downward in reality. No, the glaciers were propelled by their own weight, or more to the point, the downward pressure exerted by packed snow and ice piled thousands of feet thick.

To get a crude idea of how this works, think about taking a bucket of sand and very slowly pouring it out over a point on the ground. You get a cone-shaped pile that, as it gets higher, gradually spreads out at the base. Glacial motion is not the same, but it's similar in that the base is what spreads out — that's the source of glacial movement. It's not as if the entire mass of the glaciers slid down from Canada. Instead, immense weight and pressure caused the glaciers to spread outward in all directions from their centers.

The glaciers that sculpted our land formed over many years of cold weather as snow fell faster than it could melt. As it accumulated, the snow became compressed, changing from light crystals to hard, round pellets of ice. As more snow fell and the weight of it increased, those pellets changed to a dense, grainy ice called firn. Year after year, multiple layers of this firn built up. When the ice reached a certain thickness, it began to move under its own weight. In a process known as compression melting, the glacier applied so much pressure that the firn and snow melted even as the temperature stayed very cold. This meltwater in effect lubricated the bottom of the glacier and enabled it to spread across the landscape.

The glaciers mostly moved at the rate of only a couple of inches per day, but they did so with absolutely incredible power, crushing and grinding everything in their path. When the glaciers melted, our lakes were left behind—gifts of the ice age that have endured for thousands of years.

HOW PRECIOUS IS WATER?

A saying goes, "They're not making any more waterfront property." Another thing they're not making more of is water itself. Lay aside its beauty and its recreational values—water is absolutely essential to our lives. Water makes up about two-thirds of our bodies. A person can live for about a month without food but only about a week without water. The animals and plants we eat also rely on water.

Most of the water we enjoy today was already on our planet many millions of years ago. We drink, wash with, boat upon, and play in the same water that helped make up the tissues of dinosaurs. It's here on Earth to stay; it just keeps getting recycled through evaporation and rainfall. We can't destroy it or use it up. What we can do is keep it clean, or make it dirty, or choke it up with weeds.

As much water as there is on earth—more than 300 million cubic miles—most of it isn't really available to us. The oceans cover nearly three-fourths of the planet's surface, but they are salty and undrinkable (although they do provide a means of transportation and a sense of wonder and produce fish and other seafood for our dinner tables). Of all the water on earth, about 97 percent is saltwater. Of the remainder that is freshwater, about 68 percent is trapped in glaciers, including those in Antarctica. The National Geographic Society estimates that, in total, only about 0.007 percent of the planet's water is available for drinking and to grow crops and livestock for Earth's 6.8 billion people.

One problem with the global water supply is that it is not equally distributed. The Great Lakes, for example, are enormous water sources confined to a relatively small geography. In many areas of Earth, water scarcity is a growing concern. Eighty-five percent of the world's people live on the driest half of the planet. If there's a lesson here, it's that we who make our homes in lake-rich regions are fortunate to live amid such an abundance of water.

IT'S ALL CONNECTED

We tend to think of lakes and groundwater as separate entities. The truth is they are intimately connected. In fact, lakes, streams, rivers, groundwater—and rainfall—are all part of one interconnected water system. Groundwater is often thought of as a deep-down lake or river. In reality, groundwater fills the spaces between soil particles or the pores and fissures in rock formations. If you fill a glass with sand and then add water until all the sand is saturated, that's what groundwater is like, except that groundwater actually does flow, though very slowly.

To understand the interplay among water's different manifestations, think first about water that falls from the sky. Imagine that a landscape receives about 30 inches of rain and snow per year. Of that amount, suppose that plants and trees take up about 20 inches and release it back to the air as water vapor. The rest, about 10 inches, cycles through our water system. It runs off the land into lakes or streams or percolates down through the soil to the groundwater, the upper surface of which is called the water table. Many streams ultimately flow into lakes, but some lakes receive most of their water from the groundwater. In an important sense, a lake is a depression in the land that intersects and exposes the water table.

Many lakes send water out again through streams. Meanwhile, far below the Earth's surface, groundwater gradually moves from higher to lower elevations. The water table is not level—it slopes ever so slightly down in the direction of its movement. Through the year, rain and snow add new water to the system. The groundwater then drains out into streams and springs at the rate of about 5 gallons per second for every 1 square mile of landscape.

Of course, not every year brings exactly the same amount of precipitation. In dry years, there is less new water to recharge the system and the level of the water table will fall. In wet years, just the opposite happens. We see this in our lakes, sometimes dramatically, through extended periods of drought and in years of heavy rainfall. Rain itself (and snow) also contribute directly to our lakes. It turns out, though, that about the same amount that falls evaporates off the surface during the months of open water.

As for groundwater, never underestimate its sheer volume. To illustrate, if the groundwater under Wisconsin were a lake, it would cover the entire state to a depth of about 100 feet. Scientists estimate that groundwater reserves in the United States comprise at least 33,000 trillion gallons, about as much water as the Mississippi River has emptied into the Gulf of Mexico in the past two hundred years. The amount of water underground is twenty to thirty times as much as in all our nation's lakes, streams, and rivers.

THE IMPACT OF LANDSCAPE POSITION

Just across the road from Birch Lake where I live lies Sand Lake, less than one-fourth the size of Birch, much clearer and, unlike Birch, free of autumn algae blooms. On the other hand, Sand Lake contains mainly largemouth bass and bluegills, while Birch Lake hosts bluegills, small-mouth bass, yellow perch, walleyes, northern pike, and muskies. How can two lakes so close together be so different when they lie in basically the same geology, are surrounded by the same mixed forest, and are subject to the same weather?

As things turn out, it is actually quite common for lakes just short distances apart to display wide disparities in physical, chemical, and biological properties. Researchers today attribute these differences largely to each lake's relative position in the landscape and in the regional flow of water. In my neighborhood, Sand Lake is at the highest elevation. It has no stream inlet, and so its water sources are mainly rain and snow and groundwater. Water flows out of Sand Lake by way of a small creek, which passes through Seed Lake and then into Birch Lake. By this time the water has seen more influence from surface runoff and so is richer in nutrients. Farther downstream, Bearskin and Little Bearskin lakes are progressively more fertile. From higher to lower points in the landscape, these lakes receive progressively less influence from precipitation, and more influence from groundwater and streams. It stands to reason, then, that their characteristics can differ greatly.

Researchers at the North Temperate Lakes Long-Term Ecological Research site on Trout Lake in northern Wisconsin have noted these phenomena among lakes in general and see them as important to under-standing lake dynamics. For instance, landscape position helps explain why one lake's water chemistry changes greatly after a period of drought while another lake nearby changes very little, if at all. These researchers observe that lakes positioned higher in the landscape tend to be smaller and have clearer water than those positioned lower. Lakes higher up also tend to be more potentially sensitive to the effects of acid rain and to have less diverse populations of fish and other creatures. In addition, the water levels in lakes at the top of the landscape are much more sensitive to drought—levels can change dramatically after years of dry or wet weather.

The effect of landscape position on lakes is more than just an interesting discovery—it can have major implications for efforts to manage, protect, and improve lakes. For example, knowledge of the effects of landscape position can help resource managers forecast how different lakes will respond to changes in the climate and surrounding land uses and whether and to what extent aquatic invasive species will flourish in a given lake. It is much more efficient and cost-effective to use this regional approach than to try to measure, analyze, and understand each lake individually.

YOUR LAKE IS NOT AN ISLAND

You might think about your lake as an entity unto itself, a pool of water within the landscape. In reality, your lake is part of something larger, called a watershed. The nature of the land around the lake, and the way humans use that land, can affect the water profoundly. That's because your lake is fed not just from rainfall and groundwater but also from water that runs off the land. The size of the area of land from which water feeds your lake—its watershed—can be quite small or quite large. For example, if your lake is located at a high point in the landscape, its watershed may be very limited. But if it is at a lower point and is fed by a stream, then it may be influenced by water collected over many square miles.

In the simplest sense, a watershed (sometimes called a drainage basin) is a collector of rainfall and snowfall. Scientifically speaking, a watershed includes all the lakes, streams, reservoirs, and wetlands in the drainage area, and all the groundwater underneath. In lightly populated northern regions, lake watersheds tend to be largely in a natural state. Therefore, protecting lake quality is mostly a matter of keeping the surroundings natural—avoiding excessive development that would add nutrients and other pollutants to the water. On the other hand, lakes farther south may be fed by runoff directly from farmland, or by streams that flow through cities or villages. In these cases, correcting a water-quality problem in a lake involves a difficult process of gaining cooperation from users of land upstream to change their practices and reduce pollutant contributions.

For example, a watershed organization might be formed, representing a variety of interests—community leaders, educators, municipal officials, farmers, environmental advocates, and others—to develop a management plan. The actual work might include urban street sweeping to reduce pollutant and sediment runoff into storm sewers, education programs to encourage homeowners to clean up pet waste and use fertilizers responsibly, and incentives to farmers to follow good cropping and manure management practices that limit runoff of nutrients from their land.

The point to remember is that in many if not most cases, protecting a lake involves more than just the people who own property on its shore and use it for fishing, boating, and other recreation. It means looking at the entire watershed, and every city and village, homesite, business, farm, and other entity in that watershed has a role to play. The most successful lake protection organizations are those that adopt this expanded view and do the hard and expensive yet rewarding work of building a community of interest around protecting or improving the lake's health. It is often said that no human being is an island. No lake is either: to one degree or another, every lake is the product of the surrounding land over which rain and snowmelt flow.

HOW LAKES GET THEIR WATER

One way to classify lakes is by the way their water comes in and goes out. The number of lake types based on these criteria depends in part on who is doing the defining, but generally there are six.

Which type is your favorite lake? If you don't already know, consider doing some checking to find out.

Drainage lakes

On drainage lakes, a stream brings water in, and a stream takes water out. That is, the lake has an inlet (sometimes more than one) and an outlet. The water level in these lakes tends to stay fairly constant. Think of a bowl into which you run a slow flow of water from the tap: an equal amount of water flows in and flows out. The surface level is self-regulating.

Water Source
Drainage Lake

Eric Roell

Seepage lakes

Seepage lakes have no stream flowing in or out. Their water comes mainly from groundwater, supplemented by rainfall and runoff. The water levels are cyclical, rising and falling with wet and dry years and the related effects on the water table.

Water Source
Seepage Lake

Eric Roell

Spring lakes

The spring lake has no stream inlet but does have a stream flowing out. The lake gets its water mainly from groundwater. Many streams originate in spring lakes.

Drained lakes

Like spring lakes, drained lakes have an outlet stream but no surface inlet. They differ in that they get their water largely from rainfall, snow, and runoff, with a lesser contribution from groundwater. For that reason, their levels can fluctuate greatly. During long dry spells, the streams flowing out of these lakes can dry up.

Perched lakes

Perched lakes are truly landlocked. They have no inlet, no outlet, and no contribution from groundwater. In fact, they sit on relatively high ground, above the water table, with dense bottom sediments that hold the water in. Water levels in perched lakes can drop dramatically during drought.

Eric Roell

Reservoirs

Like drainage lakes, reservoirs have a stream inlet and outlet. The difference is that they were created by humans—they wouldn't exist if not for dams.

HOW LONG DOES YOUR LAKE'S WATER STAY?

It's easy to think of your lake as a pool of water that just sits there. In reality, water moves in and out of your lake all the time. It can enter from a creek, groundwater springs, rain runoff and snowmelt, and rain and snow falling directly into the water. It can exit through a creek, seepage into surrounding soil, and evaporation. The length of time it takes for a given drop of water that enters your lake to make its way back out is called the lake's retention time (also residence time, flushing time or water age). Depending on various factors, your lake's retention time may be a few months or several years. For perspective, Lake Superior's retention time is 191 years; Lake Michigan's is 99 years.

In general, there is no "good" or "bad" when it comes to retention time. We might think, for example, that a lake dependent on groundwater and with a long residence time would become stagnant, while a lake with an inlet and outlet stream and a short retention time would benefit from the constant infusion of new water. But if the groundwater lake is surrounded by homes with good septic systems and receives little or no contamination from fertilizer or sediment, its water quality may be excellent. And if the other lake in the example is fed by a creek subject to farm runoff or other pollution, its water quality may be very poor.

Retention time becomes important where a lake has been harmed by pollution and a lake association wants to clean it up. Here, once management practices have improved, a lake with a short retention time will tend to recover quickly. A good example is Lake Erie. Considered dead in the early 1970s, it was reborn soon after the first Clean Water Act, partly because the act dramatically cut pollutant releases but also because the lake's retention time is just 2.6 years.

As for calculating your lake's retention time, it's not easy to do with precision, but on some lakes it wouldn't be hard to get at least a rough idea. For example, if a stream flows in and out of your lake, you could use a contour map to crudely estimate the lake's average depth and, knowing its area, estimate the volume. Then you could estimate the stream's flow rate leaving the lake. From there it's an exercise in arithmetic to get to a crude measure of retention time. Of course, that calculation would ignore factors like evaporation and groundwater seepage, but it would give you at least a relative idea of how long water stays in your lake's basin.

IN THE ZONE

Scientists like to name and classify things, so it should be no surprise that they have names for different zones in our lakes. Mainly they talk about four of them. As you look out over your lake, try to picture these zones. You'll understand a little bit better how your lake ecosystem functions.

The littoral zone

The littoral zone is the area of the lake that we think most about, that we interact with the most. This is the area from the shoreline out to the point where the water is deep (or murky) enough so that there is too little light at the bottom to support rooted plants.

The width of this zone as measured from shore outward varies greatly with water color and clarity, and with the slope of the lake bottom. Where the bottom slopes down steeply, the littoral zone may be quite narrow. If the bottom slopes gradually, it may extend from shore far into the water. In fact, a shallow lake may be all littoral zone.

Life can be incredibly diverse in this zone. Most fish spend the majority of their time there, and as a consequence so do anglers. It's a rich environment, with relatively warm water, plenty of light, and nutrient-rich bottom sediments. All manner of plants grow there, from emergent species like bulrushes, cattails, and arrowhead, to floating-leaf plants like water lily, spatterdock, and watershield, to submerged vegetation like pondweeds, wild celery, and milfoils. Algae are also abundant, some species clinging to the larger plants or to rocks and sunken logs, and others floating freely. The plants provide cover for young fish, which in turn attract larger predators. Frogs, muskrats, turtles, insects, and other creatures populate this zone.

Lake Zones

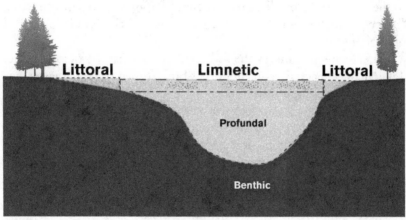

Eric Roell

The limnetic zone

Out beyond the littoral zone lies the limnetic zone, the open-water world. It begins where the littoral zone ends; its depth again depends on how far down light can penetrate. Fish move in and out of this zone, but for the most part its inhabitants are phytoplankton (algae) and zooplankton, which eat by filtering algae out of the water.

The profundal zone

Below the limnetic zone lies the profundal zone. This zone may not exist in shallower lakes. It's the deepwater region where light penetration is greatly limited. This is where dead matter from above goes to decay. It is relatively cold, dark, and oxygen-poor. The primary life in this zone consists of heterotrophs—small creatures that eat the dead material.

The benthic zone

The benthic zone, also known as the benthos, holds the lake bottom sediments. When you leave an airport, you see signs that say Ground Transportation. After flying at high speed and great altitude, above the clouds, that mode of travel seems quite unglamorous. So it is with life in the benthos. Up above in the water column the fish are like the aircraft and birds of our dry-land world. Creatures less appreciated live on and in the "ground" below. The term "benthos" comes from a Greek word, "bathys," which means "deep." It's a zone much richer in life than most of us appreciate.

Life-forms just under the sediment surface can be quite diverse. They include bacteria and fungi that break down organic matter, releasing and recycling nutrients, along with various worms and small crustaceans. Of course, crayfish live on the bottom, as do clams, mussels, and snails. Aquatic insects like mayflies, dragonflies, damselflies, and midges live on the bottom, or buried in sediment, at stages of their metamorphosis from egg, to nymph, to winged adult. These creatures are important links in the lake food chain. They eat algae or sunken plant matter and in turn provide food for fish, as anyone who has ever caught bluegills with nymphs or perch with wigglers can attest.

Leopard frogs and bullfrogs become benthos dwellers in winter. They do not (as many believe) dig into the bottom—the sediment contains too little oxygen to get them through until spring. Instead, they lie on the bottom or only partly bury themselves. Some may even swim around slowly from time to time. Painted and snapping turtles, on the other hand, do burrow into soft lake bottom mud and hibernate. In that state, they need very little oxygen and can absorb it through exposed mucous membranes in the mouth and throat.

An important function of the small benthic creatures—the worms and insects—is that they allow scientists to assess water quality in a lake. Researchers can take "grab samples" of the bottom sediment, sort out and identify the organisms it contains, and get a good idea how healthy the lake is. One measure they use is species diversity. In general, the more different creatures they find, the better the water quality. Another criterion is pollution tolerance. If a bottom sample is rich in immature forms of mayflies and stoneflies, which are sensitive to pollutants, that indicates good water quality. But if only midges and worms are present, that signals a less healthy condition.

STORIES IN SEDIMENT: PALEOLIMNOLOGY

The bottom of your lake has stories to tell. For many centuries, sediment has been collecting there, year after year, layer on layer. Materials wash in from the surrounding land. Dead matter from aquatic plants and creatures sinks out of the water column. Chemical compounds form in the water and settle out of solution. It's all there at the bottom, ready to reveal, if explored, deep secrets about your lake's history. The study of lake sediments is called paleolimnology. It's really the only way to look back in time at what a lake has gone through and how it has changed.

The water itself gives just a snapshot in time—information is available only from the years in which different parameters have been measured and analyzed. At best that has been done over a tiny slice of geologic time, and in most cases for just a fraction of the time in which modern humans have interacted with the lakes. Studying sediments back to the time of the glaciers is of considerable interest to scientists, but a more immediate aim of paleolimnology is to study what has happened to lakes since the time of human settlement around them. Such studies make it possible, for example, to determine whether changes in lakes are being caused by natural processes or human activity.

Paleolimnologists analyze the chemical, biological, and physical properties of sediments. The data they gather can be used to detect changes over time not only in the lake itself but in the land around it and the air above it. They start their studies by taking core samples from the sediment, in the same way that geologists take core samples to study the history of the Earth. The deeper paleolimnologists drill into the lake bottom, the farther back in time they travel. Glacial lakes may have sediment 15 to 30 feet deep, dating back to their formation. Once the core samples are taken, scientists use a variety of techniques to determine how old the different levels of sediment are. This includes analyzing the activity of naturally radioactive materials. Paleolimnology is one more way in which scientists can learn about lakes.

TROPHIC STATUS—ONE WAY TO CLASSIFY LAKES

There are various ways, scientific and otherwise, to classify lakes: Large versus small. Shallow versus deep. Clear water or stained. One classification matters perhaps more than the others: trophic status. That's an expression of how rich lakes are in the nutrients that support life. Typically, more nutrients—chiefly nitrogen and phosphorus—mean greater growth of algae and plants, and often by extension more fish, insects, mollusks, and other life.

Conventional wisdom has it that from the time any lake forms, it is slowly dying. It receives nutrients that feed algae and plants, which die and decompose. Nutrients proliferate until the lake gets choked with weeds; the weeds die and sink to the bottom and the lake slowly fills in. That's an overly simple description of a process called eutrophication. Scientists typically place lakes into three trophic states along a continuum of eutrophication: oligotrophic, mesotrophic, and eutrophic.

In reality, not every lake fits neatly into one of these categories—sometimes the lines get blurred. It's a value judgment to think of clearer, lower-trophic lakes as "better" than others; it all depends on how you want to use the lake. Some eutrophic lakes, for example, are terrific fish factories. Others, partly surrounded by marshes, are great spots for duck hunting or wildlife observation. As you review this list, think of where your lake fits on the trophic scale. You should know enough to take a good stab at placing it in the right category.

Oligotrophic: Deep and clear

Most lakes in northern latitudes started life as oligotrophic—poor in nutrients. They were formed from glaciers and were surrounded by infertile land, so nutrient inputs were severely limited. You can pretty well assess whether a lake is oligotrophic just from simple observations. In oligotrophic lakes, you can look down and see the bottom at a considerable depth—anglers often refer to them as "gin clear." They are tough to fish, partly because the fish can easily see their pursuers, and partly because there are not so many fish to be had. Lack of nutrients means the food chain is rather sparse. Although algae in such lakes tend to be diverse, their numbers are relatively low. Since algae form the base of the food chain, there isn't much nutrition to translate into fish flesh, although populations of large fish may be present.

The shorelines of oligotrophic lakes tend to be steep and rocky. The bottoms usually consist of clean rocks, gravel, or sand, low in organic matter and also low in sediment-dwelling organisms. Rooted plants are scarce. You tend not to see big expanses of water lilies or deep beds of cabbage weeds, as you would on lakes richer in nutrients. Since plant life is limited, there is little organic matter to decompose and consume oxygen. That means these lakes can be rich in dissolved oxygen from the surface to the bottom all year. As a result, if deep and cold enough, these lakes can support species like lake trout that depend on well-oxygenated water.

Oligotrophic lakes are undeniably beautiful and for snorkelers and scuba divers, their clear waters can be a paradise. But if fishing action is

Trophic State
Oligothrophic

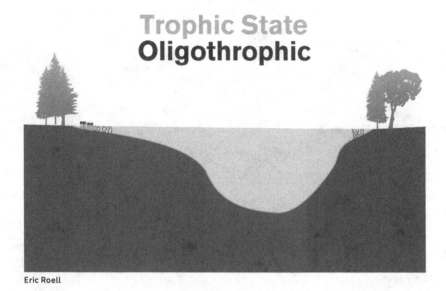

Eric Roell

what you crave, a lake higher on the nutrient scale may be more to your liking.

Eutrophic: Green and weedy

Eutrophic lakes, on the other end of the trophic scale, tend to be shallower with mucky bottoms. They may become weedy in summer, and the water may be murky from floating algae. These lakes are likely to hold warmwater fish like northern pike, bass, and bluegills. They may also be home to species like bullheads and carp that tolerate low levels of dissolved oxygen. Mention a eutrophic lake and many people will picture a stagnant pool, the water opaque and bad smelling and generally unpleasant to be around. It really isn't that simple. A eutrophic lake by definition is at a fairly advanced stage of eutrophication, but that doesn't mean the lake is "dirty" or "polluted" or otherwise undesirable—although that can be true.

The difficulty with eutrophic lakes is that they are rich in nitrogen and phosphorus, which relentlessly feed plants and algae. Blame for the nutrients often gets placed on human sources—uncontrolled stormwater runoff from city yards and streets, runoff from overfertilized farmland, poorly maintained septic systems, and others. But nutrients also come from natural sources as, for example, when a shallow lake is surrounded by and receives runoff from land with fertile soils and abundant organic

Trophic State
Eutrophic

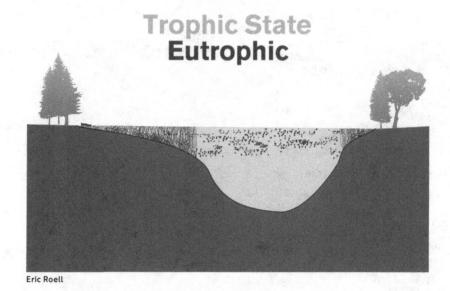

Eric Roell

matter. That is to say, some lakes are naturally eutrophic, and no amount of water-quality regulation or watershed management will change that.

Mesotrophic: Right in the middle

Mesotrophic status is to some extent the best of all worlds—it is "just right." A mesotrophic lake typically doesn't get seriously choked with weeds, nor does it normally see the late-summer algae blooms that can plague eutrophic lakes. It's not as crystal clear as an oligotrophic lake, but it is reasonably clear, enough so to allow decent snorkeling, for example, especially in June and July.

In general, mesotrophic lakes support more diverse plant, fish, and other aquatic life than lakes in the other two trophic states. You typically won't find cold-water fish like lake trout in these lakes because the deep, cold water gets depleted of oxygen by late summer. However, mesotrophic lakes can support excellent fisheries with panfish, largemouth and smallmouth bass, walleyes, northern pike, and muskies, in varying proportions.

The trick with mesotrophic lakes is keeping them that way—that is, making sure that excessive nutrients don't get in and start tipping the scale toward the eutrophic side. That means people who live on the lakes should take measures like keeping their septic systems well maintained,

Trophic State
Mesotrophic

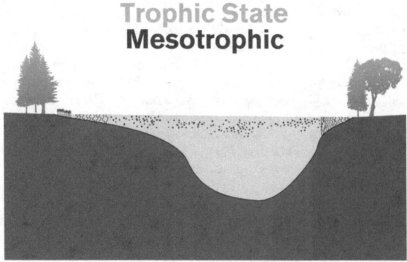

Eric Roell

avoiding excessive yard and garden fertilizers, and limiting hard surfaces that increase stormwater runoff. Actually, these are wise practices to follow on any lake.

THAT ALL-IMPORTANT OXYGEN

We know that fish and other organisms can live because of oxygen dissolved in the water. The question is how that oxygen gets into the water and stays there. First of all, oxygen doesn't "dissolve" in water in the same manner as salt or sugar. Salt, for example, is a compound of one sodium and one chlorine atom. In water, it ceases to be a solid, splitting into its parts: a positively charged sodium ion and a negatively charged chlorine ion. Oxygen gas, on the other hand, consists of two oxygen atoms, and it remains in that form when mixed with water. So instead of referring to dissolved oxygen, perhaps we should speak of suspended oxygen, or entrained oxygen.

In any case, dissolved oxygen is arguably the single most important component of water quality because, quite simply, without adequate dissolved oxygen, next to nothing lives in the water. Oxygen gets into your lake mainly from the atmosphere, which contains about 20 percent oxygen. The amount of oxygen in water is tiny compared to the amount in the air. A healthy lake or stream will contain 6 to 8 parts per million (ppm) of oxygen, or 0.0006 to 0.0008 percent. If dissolved oxygen is less than 2 or 3 ppm, most fish will become stressed or die.

Not all the oxygen in lakes comes from the air. Some is created through photosynthesis, the process by which plants turn sunlight into food. Besides the plants you easily recognize like cabbage weed, coontail, and water lily, your lake's water is full of algae. The plants and all these algae constantly photosynthesize and produce oxygen while the sun shines. But if that's true, one might ask why a lake with heavy weed growth or in the middle of an algae bloom isn't rich in oxygen, and why in fact an algae bloom can deplete oxygen and kill fish.

To answer, remember that photosynthesis requires sunlight, so it doesn't happen at night and happens much less on days with heavy cloud cover. During those times, the plants and algae consume food and use up oxygen, just like most other living things. Furthermore, the algae can die off en masse, at which point oxygen-using (aerobic) bacteria begin to break the algae cells down. That can cause dissolved oxygen to plummet to levels unsafe for fish.

So, while algae and plant life contribute to dissolved oxygen, they are a bit of a mixed bag. The main source of life-giving oxygen in your lake is the atmosphere. And the most oxygen mixes in when a healthy wind kicks up waves. So next time you see a wind raising whitecaps out on the water, imagine that your lake has just filled its lungs with a big, deep breath of fresh air. And now the fish can breathe better, too.

YOUR LAKE HAS LAYERS

While your favorite lake may appear to be a big, uniform pool, that water actually has a structure. It has layers, at least for much of the year. Your lake has layers for the simple physical reason that a less-dense liquid floats on a more-dense one, as oil floats on water. That may not be a pretty image, but it makes the point.

Imagine your lake just after ice-out. As spring advances and transitions to summer, warm air and sunlight constantly pump heat into the lake. The result is that the water near the surface gets warm, while the water down deeper stays cold. Eventually, your lake has a layer of warmer (less dense) water floating on a layer of colder (denser) water. Through summer, the layers don't mix very much. At the warmest time of year, the difference in density between surface water (at, say, 80 degrees F) and the deep water (at 45 to 50 degrees F) is considerable.

Thermal Stratification

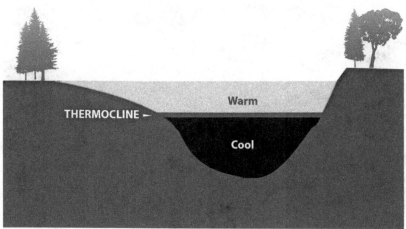

Eric Roell

The technical term for this condition is thermal stratification. The warm upper layer is the epilimnion, and the cool lower layer is the hypolimnion. Between those layers, there's a zone where the temperature changes rapidly—that is called the thermocline (some say metalimnion). On many lakes, it's easy to experience the thermocline, and you may have done so without knowing it. Next time you go swimming, get out where the water is fairly deep, say 15 feet or so. The surface water you're in will be reasonably warm. Now, do a feet-first surface dive. With an upstroke of your arms, propel yourself down. Soon your feet will feel a sudden cooling. That means they have hit and passed through the thermocline. With this simple test you can roughly determine the depth of the warm, upper layer.

Most lakes stratify, or form layers, in summer. As summer wears on, all kinds of materials sink from the warm surface water into the cold bottom layer. Plant parts, algae, fish carcasses, dead insects, and more drift down and decompose, consuming oxygen. As a result, the oxygen in the depths can become quite depleted. If your lake remained stratified all the time, those deep waters would become largely lifeless, habitable mainly to organisms that thrive in anaerobic (without oxygen) conditions. But fortunately, along comes the fall turnover, generally in late September or early October. Fall turnover is a restorative process, a bit like opening doors and windows in a long-sealed, musty cabin and letting lots of clean, fresh air course through.

Fall Turnover

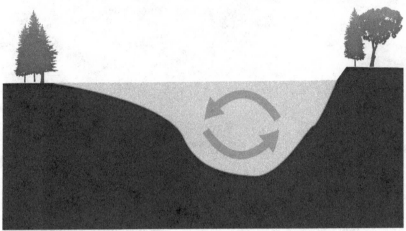

Eric Roell

As summer fades into autumn, the surface water gradually cools, and the difference in density between the surface and deeper water decreases, so that eventually wind and wave action can mix the layers together. And that means the lake, from surface to bottom, becomes infused with oxygen. This is great for all manner of lake creatures that need oxygen to make it through the winter.

It's not hard to tell when your lake is turning over. For one thing, the water suddenly becomes cloudier than usual because the mixing action brings up living and dead algae along with debris from deeper in the water column. There might even be, just for a short time, a hint of sulfur scent in the air, like rotten eggs, as decomposing material comes to the surface. When the turnover is complete, the water becomes clear again, likely more so than in high summer.

Some anglers say fishing is tougher during the turnover because with oxygen available everywhere, the fish are more scattered. Lakes experience fall turnover in different ways. Deeper lakes take longer to turn over. Shallow lakes may not turn over at all because they never actually stratified in the first place—wind and wave action keeps them well mixed all through summer. The turnover itself can play out in a few days in some lakes or a week or more in others. Fall turnover is a seasonal milestone, like ice-in and ice-out, that can be fun to track over the years.

With the onset of winter, the lake once again becomes layered up as the water becomes very cold. The process is the reverse of what

Thin Soup

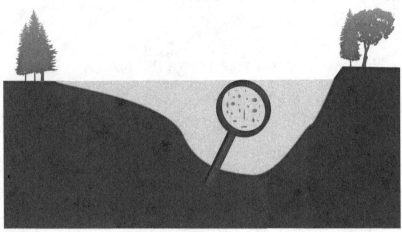

Eric Roell

happens in summer. The relatively warm water at 39 degrees F, being the densest, sinks to the bottom. The coldest water—the least dense— remains on top. Once again, your lake is layered up. And as the weather warms again, the cycle repeats itself.

THIN SOUP

One way to think of your lake is as a thin soup of meat and vegetables. The water is the broth, and the meat and vegetables are planktons of two basic kinds: algae (phytoplankton) and animals (zooplankton). Plankton is an incredibly rich and important source of food in a lake's ecosystem. Under a microscope, the little zooplankton creatures are fascinating to observe and the algae are amazing in their variety, symmetry, and, yes, beauty.

Your lake's water is loaded with plankton. Just how loaded depends on where your lake falls on the nutrient scale—from oligotrophic to eutrophic. When you look down into your lake, you are looking not just at water but through a liquid full of life-giving food. Plankton forms the base of any lake's food chain, or perhaps more correctly, the anchor points for the lake's food web. It's plankton that tiny fish fry eat when they are too small for anything else. Bigger fish then eat the fry and in turn get eaten by still bigger fish, and so on all the way up to the mighty muskellunge. So, in the end, no plankton, no musky. If big oaks grow from tiny acorns, then in a somewhat different sense, trophy fish grow from tiny plankton.

IT ALL STARTS WITH THE SUN

For the walleyes, bass, and pike we catch from our lakes, we can thank the smaller fish they eat. For those prey fish we can thank the smaller organisms they consume (insects, animal plankton, algae). And for all of it we can thank the sun, because that's where everything begins. Yes, the sun drives the process in our lakes that's called primary production, a fancy way to say "making food." Plants, most notably algae, are a lake's primary producers because they receive sunlight and create food through photosynthesis.

Photosynthesis, which occurs inside individual plant cells, is a chemical process in which carbon dioxide and water are converted to oxygen and carbohydrate. This carbohydrate is the energy source that sets the food chain in motion, all the way up to predator fish and, ultimately, to us. Of course, there's more to the creation of plant life than carbon dioxide and water. Algae and water plants also depend on the primary nutrients nitrogen, phosphorus, and potassium, and smaller amounts of other nutrients including calcium, sulfur, magnesium, iron, zinc, and copper.

The abundance or absence of algae and plants depends on factors including water depth, bottom type (muck, sand, gravel, rock), and water clarity (which determines how deep sunlight can penetrate). The key driving factor, though, is the relative abundance of the major nutrients. If just one nutrient is in short supply, then plant growth is restricted. In our lakes, algae and water plants require forty-one parts carbon, seven parts nitrogen, and one part phosphorus in order to grow.

Next time you enjoy a meal of fish from your lake, remember the interplay of nutrients and the primary producers, and give a little thanks to the algae and the sun.

HOW CLEAR IS YOUR LAKE?

Is your lake clearer than the next one over? Clearer than the one where your best friends live? Maybe the questions don't matter to you, but water clarity is one attribute people think of when judging the quality of a lake. One way to determine whose lake is clearer is to buy (or make) a Secchi disc, a simple tool for measuring water clarity with reasonable consistency. It's quite a bit of fun to use. The Secchi disc (pronounced SEKK-ee) is a standard device lake scientists use. Look at a Department of Natural Resources report on fish stocking in a lake and you're likely to find a Secchi disc reading.

A Secchi disc is a circular, platelike object painted with quadrants alternating white and black and suspended on a rope. To use the disc, you go out in a boat to where the water is the deepest, then slowly lower the disc until you can no longer see it. Mark the rope at the waterline with a clip clothespin. Then slowly, slowly, pull the disc up until you can see it again. Once more, mark the rope at the waterline. Now retrieve the disc and measure the distance from the disc to both marks. Take the average, and that's your Secchi disc reading.

What the Secchi disc actually measures is how deep light penetrates into the water. Suppose your Secchi disc disappears at 10 feet below the surface. It would seem that means the light penetrates to 10 feet. In reality, the light you see bouncing back from the disc actually has to make a round trip, from the surface down, and then back up to your eyes—20 feet in all. If you take that Secchi disc reading of 10 feet and multiply it by 1.7, that gives you the maximum depth (17 feet in this case) at which there is enough light for photosynthesis in plants to occur.

A Secchi disc is quite easy to make. You can search on the internet for instructions on how to make one suitable to your purposes. All you really need is a round, flat object heavy enough to sink; a drill to bore a hole in it; a length of rope; and cans of white and black waterproof paint. As long as you can get the disc to stay faceup in the water, it will do the job. Try making one and checking your lake water clarity, not just once but at different times of the year, including winter, through the ice. Keep a record of Secchi disc readings in your home or cabin log. It can be interesting to track the annual and seasonal trends.

THE COLOR OF WATER

What color is your lake? Actually, that depends—on from where you look at it, on what's happening with the weather, and on what's in the water. Let's start with that blue we most often associate with lakes. Some say the water looks blue because it reflects the sky, but that isn't so. The blue of lake water is a deeper, richer shade than the robin's-egg blue of the heavens. Scientists note that water itself has a very faint blue tint, caused by nuclear motions (excited vibrations) of the water molecules, but there's no need to go to that level of complexity.

Your lake looks blue because of the way light behaves after it enters the water. Sunlight includes the seven colors of the visible spectrum: red, orange, yellow, green, blue, indigo, violet—the colors of the rainbow. Light consists of waves, and each color has a different wavelength. When the sun shines on water, all colors of light penetrate the surface. The colors with the longest wavelengths—red, orange, yellow, green—are absorbed in the water. That leaves mostly the shorter-wavelength blue light to reflect back to our eyes.

Of course, our lakes don't always look blue. Under the dark clouds of a storm, for instance, the surface may take on an ominous steel gray. On a day with white sky and a light fog, the water may appear whitish or light gray. In these cases, the water does take on the sky's appearance. Lakes can also reflect the beautiful hues of sunrises and sunsets.

Blue is the color we see if the water is reasonably clear. Particles in the water influence its color. If you've observed a Great Lake on a sunny day, you may have noticed that the water looks tan near the shore, greenish blue farther out, and then deep blue to the horizon. The tan is the color of the bottom sand, particles of which are suspended in the water, especially if waves are breaking. The greenish band is caused by abundant free-floating algae. The deep blue comes from those selectively reflected blue light waves.

We see the blue and other colors mostly when looking at the water from some distance. If we gaze straight down into the shallows, as from a pier, the water most likely looks colorless, the same as when we look into a glass or pitcher of water. If your lake is in the midst of an algae bloom, it will appear pea-soup green, whether you're looking downward or out across the surface.

If you observe a brownish or reddish tint in the water, your lake is high in tannins.

Lakes with this kind of coloration are often called stained. The depth of coloration varies. Some lakes are barely tinted. Others are more deeply colored, like tea, or even coffee. Looking at bottles of water taken from a clear lake, a lightly stained lake and a deeply stained lake against a white background, you would see just a slight difference in color. The difference really appears when you look down through two or three feet of water over a sand bottom.

The color comes from a process that is a lot like brewing tea. The brown or red comes from tannic acid, a dissolved organic chemical that leaches into the water as wood, leaves, and other vegetation decompose. It may help to think of places like bogs, wetlands, and swamps as analogous to tea bags. You could observe the effect by putting some leaves in a bucket of water and letting it sit for a month or two. Lakes near wetlands and bogs or surrounded by coniferous woods are the most likely to be tannin-stained. So are lakes heavily influenced by rivers, which constantly pick up staining from vegetation as they collect runoff from their surroundings. This explains why many flowages have stained water.

Colored water doesn't mean a lake is impaired. In fact, many stained lakes (also called dystrophic lakes) have excellent water quality. But the coloration does affect lake life. For one thing, aquatic creatures, from insects to fish, tend to take on darker coloration in stained water. Also, anglers know that certain colors of lures and jigs attract fish more effectively in stained water than clear.

Sunlight does not penetrate as deep in stained lakes, as the color absorbs the light. This means plant growth is typically less dense and is limited to shallower depths than in clear water. Stained lakes also tend to be slower to warm up in springtime. There's a kind of wild beauty to stained water. It adds another level of diversity to the lake country we call home.

LIGHT IN THE WATER

Light is an essential life force in lakes. How it behaves in the water depends on various factors, some immutable—pure physics—and others related to the characteristics of each lake. Let's start with one of those immutable behaviors—refraction. Light (and here we mean sunlight) slows down when it strikes the water, and as a result, the direction in which the light waves travel bends upward. So, imagine that you are an archer and see a fish in shallow water. Because of refraction, the fish appears higher in the water column than it actually is. So in order to hit the fish with an arrow, you would have to aim slightly below where you see the fish's image.

Another immutable property is reflection. The percentage of the light meeting the water that reflects off the surface depends on the angle at which the light travels. At noon on a clear day, when the sun is almost directly overhead, about 95 percent of sunlight penetrates into the water—just 5 percent or so is reflected back. But by late afternoon when the sun is lower, about half the light is reflected.

As for the light that penetrates the surface and enters the water, that behaves differently depending on what the lake and its water are like. In general, portions of the light entering the lake are absorbed by plants to make food by photosynthesis, absorbed by the water or the particles in it and converted to heat, scattered back out of the lake by floating (suspended) particles, and reflected or absorbed and converted to heat by the bottom sediments.

The character of the lake determines how much of the light behaves in these ways. For example, dark bottom sediments absorb more light (and heat) than lighter-colored sediments. Water with a high concentration of suspended particles may absorb and reflect most of the light within the first few feet from the surface, so that plants can't get enough light to grow except in the shallowest areas. The dissolved substances in stained water have a similar effect.

Naturally, snow and ice cover in winter reduces light penetration. A layer of crystal-clear ice transmits light just about as well as clear water, but if the ice is cloudy because of air bubbles or impurities, or is stained by organic matter, it absorbs more light and allows less to pass through. Snow increases sunlight reflection off the surface by about 75 percent. When the snow is heavy, very little light penetrates the surface. So, as the sun sinks lower with the approach of winter, as the light strikes the water at a sharper angle, and as ice and snow cover take hold, the lake becomes a progressively darker place.

Light reflection also affects how well we can see into our lakes. Many of us know we can discern more features in shallow lake water when wearing polarized sunglasses. It's easier to see fish we want to catch, observe bottom characteristics, or discover treasures like lost fishing lures with those glasses on. Polarized lenses work by filtering out reflected light. Light waves are oriented in all directions. Light waves that reflect off water (or any surface) are oriented horizontally—think of a sheet of paper, parallel to the water, coming toward you edgewise. The lenses of polarized sunglasses are specially treated so as to form a filter that acts like a vertical picket fence that stops those horizontal waves, except that the spaces between the fence "slats" are extremely small.

Imagine you're holding a long rope that's tied off against a tree. Between you and the tree is a picket fence, and the rope passes between two of the fence slats. If you were to move your end of the rope rapidly up and down, you would create a wave in the rope, and that wave would pass right between the fence slats and reach the tree. Now imagine moving the rope rapidly side-to-side. The narrow space between the slats would block the formation of the wave, which would never reach the tree.

That's what polarized sunglass lenses do to reflected light. Light waves oriented horizontally that would otherwise impede your ability to see into the water are filtered out before they hit your eyes. Light waves with a vertical orientation are allowed to pass through, revealing all those secrets from below the surface. If you wear your polarized glasses while exploring your lake, you'll get to know a little more about what lies below, though not as much as if you looked through a swim mask.

WAVE ENERGY

As a little kid growing up on Lake Michigan, I would sometimes take a stick of driftwood from the beach, throw it out into the water, and wait for the waves to bring it back. I wondered why it took so long, if indeed it ever came back at all. There's a reason it didn't return. Our perception of waves as water rolling toward shore is a great and wonderful illusion. What we really see in waves is energy rolling beneath the water's surface. The water itself doesn't travel.

Consider this analogy: You and a friend hold the ends of a long rope and move your arms up and down in the same rhythm. Waves will develop that travel the length of the rope, but the rope fibers don't move horizontally—only up and down. That's what happens with waves. A surfer on the ocean riding a wave is actually propelled by pure energy. The water is only the substance through which the energy flows. A free-floating object caught in the waves travels in vertical circles, essentially bobbing in place. It makes progress only when a whitecap catches it and pushes it along.

Of course, we don't see the size of waves on our inland lakes that we observe on the ocean or the Great Lakes, except perhaps in the teeth of an especially violent storm. Still, it's interesting how waves form and behave. Start by picturing a mirror-smooth surface on your lake and you can mentally watch how ripples, then waves, then whitecaps are created. First a breeze stirs up small ripples. The sloping surfaces of these ripples catch more wind. As the wind gathers strength, the ripples build into wavelets, then into larger waves. The size of the waves depends on three main factors: the speed of the wind, the length of time the wind blows, and how far the wind blows across the water without hitting obstacles (the fetch). The greater each of these variables becomes, the larger the waves.

As the wind blows toward shore, it pushes the waves ahead. They roll along until they reach shallow water. Contact with the lake bottom slows the lower part of the wave down; the top maintains speed and so tumbles forward and collapses. That causes whitecaps. It's as if the wave gets tripped, as we do if we hit a taut ankle-high rope while running. Of course, waves can whitecap out in the open water too; that happens anytime the top of the wave becomes too large and heavy for its base to support. For measuring the size of waves, there are four variables:

- Height: The distance from the bottom of the wave (trough) to the top (crest).
- Length: The distance between crests.
- Steepness: The angle between the trough and the crest.
- Period: The length of time in the wave's movement between one crest and the next.

One of the pleasures of lake life is quietly watching and listening to the waves. It's a special form of energy that promotes such rest and peacefulness.

YOUR LAKE HAS A "SKIN"

You've seen the rounded shape of droplets on a lakeside leaf or pier board. You've noticed how it's possible to fill a glass with water to just very slightly above the brim if you keep the glass very level and very still. In both cases, surface tension is the phenomenon responsible. Because of surface tension, water behaves as if it had a (very thin) skin. In reality it doesn't; what we see with surface tension is the result of molecular forces, specifically the cohesive force that tends to hold the water molecules together.

Water molecules under the surface are surrounded by other identical molecules, and so the forces they exert on one another cancel out. But water molecules on the surface don't have others above them to bond with, and therefore they bond more strongly with the molecules next to and below them. These bonds resist being stretched or broken. To appreciate this, it may help to imagine yourself trying to push your way

Paul Skawinski, University of Wisconsin–Stevens Point

Fishing spider, showing surface tension

through a line of strong men, hands joined, arms extended. Surface tension is a little bit like that.

There are various ways to experience surface tension. Believe it or not, it is possible to (very carefully) place a small needle on water and have it float, even though its material, steel, is several times as dense as the water. That's surface tension at work. When camping in the rain as a kid in a canvas tent, perhaps your parents warned you not to touch the tent. Touching the fabric would cause water to leak through. As rain collects on the tent, surface tension in the water bridges the pores of the material, in effect creating a rain barrier. Touching the material breaks that surface tension. Drip, drip, drip.

Detergents in effect "make water wetter" (as an old commercial claimed), by reducing surface tension so that the water soaks into fabrics more easily. Similarly, hot water is better for washing because it has lower surface tension than cold water. On calm days around your pier, look for visual signs of this property of water. You may find one in an incredibly cool insect called the water strider, whose "feet" float on the surface tension.

HOW ACID OR ALKALINE IS YOUR LAKE?

A characteristic you can't see or feel can have meaningful effects on life in your lake. It's called pH, and it's a measure of how acid or alkaline your lake's water is. We know that water molecules contain two atoms of hydrogen and one atom of oxygen (H_2O). However, some of those molecules actually exist as positively charged hydrogen ions (H^+) and negatively charged hydroxide ions (OH^-). In pure water, those ions exist in essentially equal numbers. But when chemicals are added to water, the balance can shift in one direction or the other. A solution with more hydrogen ions is acidic; a solution with more hydroxide ions is alkaline (or basic).

The level of pH is measured on a scale from 0 (extremely acidic) to 14 (extremely alkaline). Pure water, which is considered neutral, has a pH of 7. Relating this to common substances, lemon juice is a fairly strong acid (pH just over 2), while household ammonia is strongly alkaline (pH about 12). Lake waters are never that strongly acidic or alkaline. Their pH falls generally in a range from 6 to 8, close to neutral.

The pH of your lake's water helps determines how well certain fish species, plants, insects, and other life-forms survive and reproduce. For example, at pH below 6.5, walleye spawning is inhibited, and small-mouth bass disappear below pH 5.5. The pH level can also determine the extent to which certain pollutants are released into the water from sediments in the lake bottom. In healthy lakes, the effects of pH levels on lake life are mostly subtle.

Lake systems have ways of regulating their pH level and neutralizing acids that are introduced. Though we tend to think of rain as pure and innocent, rainwater is acidic enough that many or most fish could not reproduce if they had to live in it. On the way down, rain collects carbon dioxide from the atmosphere, and that gets converted to carbonic acid, the same thing that gives soda pop or sparkling water its fizz. Natural rainfall unaffected by any air pollutants comes in at about pH 5.6, somewhat less acidic than black coffee. Rain that picks up air pollutants like sulfur and nitrogen oxides will have a much lower pH of 5.0 to 4.5—about as acidic as lemon-lime soda pop, or even lower.

Though the purest natural rainfall is acidic enough by itself to create a hostile environment for fish and other creatures, life is able to persist in our lakes because most lake systems include chemical substances that continuously neutralize, or buffer, the acids. Imagine that your stomach had a never-ending supply of Tums or some other antacid, dosed automatically. It's a bit like that, only a good deal more complicated.

The buffering is provided by carbonate and bicarbonate ions. Of the two, carbonate is twice as effective as a buffer. The main source of carbonate in most lakes is limestone, for which the chemical name is calcium carbonate. In water, a small amount of this material dissolves into calcium and carbonate ions. The negatively charged carbonate ion is then free to react with the positively charged hydrogen ions that make the water acidic, taking them out of circulation and raising the pH. It follows, then, that a lake's capacity to buffer acid depends on the kinds of minerals in the lake sediment and the bedrock in the watershed, and to what extent the water comes in contact with and dissolves minerals with buffering ability.

The buffering capacity will be high, for example, if much of the lake's water comes from groundwater that contains limestone. On the other hand, if the water source for the lake passes through sand made up of quartz or other minerals that don't readily dissolve, then buffering capacity will be low and the lake will tend to be acidic. This is all the more true if the lake gets much of its water from rainfall, rather than from groundwater or an incoming stream. Many lakes in northern latitudes are low in alkalinity because the glacial deposits in which they rest don't contain much limestone. For that reason they are also relatively vulnerable to acid rain.

Generally speaking, lakes high in alkalinity are more hospitable to fish and aquatic plants than those with more acidic water. The interplay between acidity and buffers is important. For the most part, it works in favor of promoting abundant life in our lakes.

THE NITROGEN CYCLE

Nitrogen is one of the three primary plant nutrients (with phosphorus and potassium). It also makes up about 78 percent of the atmosphere. Like oxygen, nitrogen dissolves in the water of our lakes. So if nitrogen is so abundant, much more so than oxygen, it's easy to wonder why all our lakes aren't overfertilized and loaded with algae and weeds. A major reason is that nitrogen in the form that exists in the atmosphere can't be used by plants.

The form in the air is called molecular nitrogen and consists of two nitrogen atoms (N_2), just as the oxygen we breathe consists of two oxygen atoms (O_2). Before plants can use nitrogen, the N_2 has to be converted to other forms in a process called nitrogen fixation. This takes a substantial amount of energy. One way this happens is through the energy released by lightning. The jolt of electricity and heat causes the N_2 to combine with water (H_2O) to form ammonium (NH_4^+) and nitrate (NO_3^-). The ammonium and nitrate then fall to the ground with rain and enter our lakes. These forms can be taken up by plants.

Lightning accounts for a fairly small amount of nitrogen fixation — about 90 percent of that work is done by bacteria, with the help of molecules called enzymes that accelerate chemical reactions. These bacteria convert N_2 into ammonia (NH_3) and ammonium. In a process called nitrification, other bacteria then convert the ammonium into nitrite (NO_2^-), and a third type of bacteria then convert the nitrite into nitrate (NO_3^-).

Nitrogen is continuously converted among these different forms in what is known as the nitrogen cycle. The process by which lake weeds and algae take up nitrogen to build their tissues also works in reverse. When the plants and algae die and decompose, bacteria convert the nitrogen in the tissues to ammonium and ammonia (ammonification). Then still other bacteria convert ammonia into nitrogen gas (denitrification). This nitrogen (N_2) can remain dissolved in the water or escape into the atmosphere. And so the cycle is complete.

HOW PHOSPHORUS BEHAVES

We're often told about the harm too much phosphorus can do to our lakes. Just a little bit can go a long way in a bad direction: one pound of phosphorus can promote the growth of up to 500 pounds of algae. As we think about how to prevent phosphorus pollution, it helps to understand how phosphorus behaves in the soil and in the lakes, and why control of runoff is such an important part of prevention.

In most of our inland lakes, phosphorus is called the limiting nutrient — the scarcity or abundance of it determines the extent to which weeds and algae grow. Imagine a long line of folks at a restaurant waiting for their Friday fish fries. The restaurant has two hundred perch filets, fifty orders of french fries, and twenty loaves of rye bread, enough for everyone — but only one small bowl of coleslaw. That means only a few diners can get a complete meal. Phosphorus in the lake is like the coleslaw. In short supply it holds plant growth back, but make it abundant and those hungry plants and algae have a feast.

Phosphorus exists naturally in the soil around our lakes, the vast majority of it in fixed forms that don't dissolve in water and can stay put for years. But another amount, about one-tenth as much, exists in an active form, attached to particles of soil and quite easily released into solution in the surrounding water. Runoff from properties around a lake will to some degree erode the soil, carrying soil particles into the water. Soil phosphorus is connected more closely with fine particles than with coarse particles and, as we might expect, finer (lighter) particles are more easily carried by water.

So, when soil erodes, more fine particles are removed than coarse particles. Therefore, the sediment leaving the soil through erosion tends to be, relatively speaking, rich in phosphorus. Soils with fine texture—those high in clay and silt particles—have the most capacity to add phosphorus to a lake if eroded soil is not kept on the landscape.

Phosphorus dissolved in water—not attached to soil particles—can also be a meaningful source of nutrient enrichment in a lake, but the amount is small relative to that carried by particles of soil. Comparatively little phosphorus enters groundwater, and ultimately the lakes, by leaching through the soil. The addition of phosphorus comes mostly from flows of water on the surface.

The behavior of phosphorus once it enters a lake is complicated. Phosphorus in soluble form is quickly used by plants and algae, but phosphorus can also react with calcium or iron to form particles that settle to the bottom (precipitate) and lie dormant. Some lakes are more sensitive to phosphorus than others. For example, in small lakes, light phosphorus loads may quickly accelerate algae growth, while deeper lakes where incoming soil and nutrients can easily settle out may handle heavier loads without much negative effect.

In the end, all of this speaks to the importance of limiting soil erosion and runoff into the lakes. All year, lake properties receive phosphorus from leaves, grass clippings, pet waste, bird droppings, airborne dust, and other sources. Phosphorus can run off from absorbent (pervious) surfaces like lawns, flower beds, gardens, and wooded areas, but runoff from impervious surfaces like sidewalks, driveways, and rooftops tends to flow faster and is much more likely to find its way to the lake, carrying phosphorus with it.

The trick is to minimize those flows or, failing that, slow them down. Steep slopes, soils compacted by construction, and sparse grass or other vegetation tend to favor more rapid runoff. Dense grass or other plantings tend to slow the flow and let the water soak in. Reducing runoff is a concept to consider as we look to maintain and improve our properties.

HOW TO KNOW YOUR LAKE BETTER THAN ALMOST EVERYONE

For getting to know your lake, there's nothing quite like a swim mask, flippers, and snorkel. I love slipping them on and exploring up and down the shoreline from our cabin on Birch Lake. While often I've taken my snorkel gear to other lakes with extremely clear water, I've also kicked my way around Birch. One great thing about snorkeling is that the action of the flippers keeps you buoyant with little exertion. That means you can venture fairly long distances from shore without worrying that you may get too tired to make it back. (Some people prefer to wear a life vest when snorkeling over deep water, just in case.)

I enjoy the new scenery I encounter while snorkeling, and I almost always learn something new about my lake. Last year I learned that our rusty crayfish population was bigger than I imagined just from what I saw around my pier. While snorkeling I saw them everywhere, some of them huge. I also noted some patches of cabbage weeds down the shoreline from our place—a good sign after the cabbage beds were all but wiped out years ago when the crayfish population exploded.

The best thing I learned, though, was the exact nature of the spot a short distance from my pier where a brother and I have caught numerous walleyes at times. About 50 yards down the shore and about 50 feet out from a reed bed lies a tangle of brush, probably placed there deliberately as a crib years ago. Hovering over it, I looked down on a couple of smallmouth bass in the upper branches and, down deeper, a dozen or more walleyes, finning in place. They seemed not to notice as I quietly passed over.

On one memorable August day, I snorkeled my way over to the edge of a shallow reef, where I stopped and stood in chest-deep water to clear fog from my mask. When I looked down again through the mask, five smallmouth bass were swimming around and between my bare legs. They stayed there for a long time, as if my legs were sunken logs or some other form of comforting cover.

I'm surprised that I rarely see anyone else snorkeling on Birch Lake, or on other lakes. You could certainly do worse than to spend some time getting to know your lake from a whole new perspective, with flippers on your feet and your face behind glass.

Paul Skawinski, University of Wisconsin–Stevens Point

FINS

·2·

FISH ANATOMY: HOW THEY'RE BUILT

Swim bladder: Benign inflation

You've surely seen fish in your lake placidly finning in place, holding at a constant depth, exerting almost no energy. That is only possible because of an ingenious feature of fish anatomy. The bodies of fish are slightly heavier (denser) than water, which means that without some effort they would sink if they held still. It's their swim bladder that enables them to achieve neutral buoyancy.

The swim bladder is an elongated bag that rests in the fish's body cavity just above the innards. Fish can fill this bladder with gas, mainly nitrogen and oxygen, to varying extents depending on the depth where they swim. Some kinds of fish, like sharks and rays, do not have swim bladders. That means they either have to keep moving constantly to stay suspended at a certain depth, or stop moving and sink to the sea floor for a rest.

In nature, energy is at a premium. Swimming or using fins continuously to create lift takes substantial energy. Thus fish that developed swim bladders and became neutrally buoyant had an evolutionary advantage. They could eat less to sustain their energy, and they had more freedom to use their fins to make intricate maneuvers and, yes, to hover in place. Fish that evolved relatively recently have swim bladders that originated as an extension of the gut. In some fish the bladder actually connects to the gut. These fish can swallow air at the surface, then force it from the gut into the bladder and thus control its degree of inflation. In other fish, the bladder gets filled in a manner more similar to the function of fishes' gills or humans' lungs. That is, gases move to the bladder by way of the bloodstream. The degree of inflation is controlled by areas where the bladder membrane is very thin and is rich in tiny blood vessels called capillaries. Gases pass from the blood cells, through the walls of the capillaries, and through the bladder membrane to the inside.

Fish control the pressure in the bladder according to the depth of the water. The water pressure on a fish's body at 30 feet deep, for example, is almost twice the pressure at the surface. The greater the depth, the higher the gas pressure in the swim bladder. Anglers who have fished for deep-dwelling species like lake trout may have seen one consequence of this. If a fish is quickly brought to the surface from great depth, the high pressure inside the bladder may cause it to expand, fill the throat, and even protrude from the mouth.

When left to themselves, however, fish can exert a fine degree of control over the swim bladder and thus their bodies' buoyancy. The swim bladder is one mechanism behind the beauty, diversity, and vitality of fish in our lakes.

The functions of fins

The fins on flashy 1950s cars were, practically speaking, useless appendages, all style and no substance. The fins on fish are a different matter. Fins add to fishes' beauty—take for example the sail on a marlin—but they are also highly functional, as essential as arms and legs are to humans. Let's start with the tail fin, or what biologists call the caudal fin. It's mainly for propulsion. If the fish's muscles are the engine, then the tail fin is the propeller. A few whips of the tail and the fish can be off in a flash, chasing prey or fleeing a predator. The tail fin also contributes to steering.

The dorsal fin, the one that runs along the top of the back, adds stability during travel, a little bit like the centerboard on a sailboat or the fletching an arrow. In many fish the dorsal fin also contains spines. It can lie flat or be unfurled, spines vertical. The expanded dorsal fin can

Typical Fin Structure

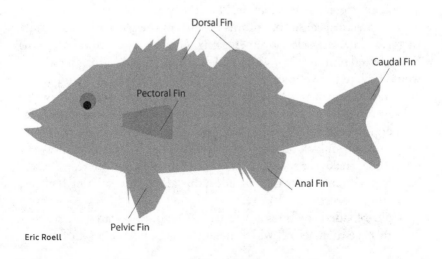

Dorsal Fin

Caudal Fin

Pectoral Fin

Anal Fin

Pelvic Fin

Eric Roell

help protect a fish by making it appear larger than it actually is to a predator. The spines themselves can also deter attacks. If you've ever been "speared" while unhooking a perch, imagine how that would feel to the inside of a larger fish's mouth.

The anal fin, on the fish's underside forward of the tail, also lends stability. Deep-bodied fish, like bluegills and crappies, generally have larger dorsal and anal fins because they need more support to hold themselves upright. Pelvic fins, on the underside coming off the belly, also help the fish stay level, providing stability against rolling from side to side. The pectoral fins, generally on the sides of the fish just behind the gills, add stability and help the fish maneuver and control depth. Pelvic and pectoral fins also act as brakes: when flared out they help the fish slow down and stop. Fin sizes, shapes, and configurations vary with fish species, where they live, and how they feed. No matter their form, they're essential mechanisms of mobility.

In cold blood

We'll never understand what it's like to live under lake ice. We'll never know the sensations because we're not fish, and we can never relate to their experience because our metabolism is radically different from theirs. We are warm-blooded; fish (along with frogs, crayfish, turtles, and other water creatures) are cold-blooded. Our bodies regulate our temperature; much of the energy we derive from food goes to fuel the inner furnace that keeps us at about 98.6 degrees F. Cold-blooded creatures' temperatures rise and fall with the temperature of their environment. That means in winter the fish in your lake are at somewhere around 40 degrees F or colder, just like the water.

In a practical sense that means the fish in the winter lake are quite sluggish. Their muscle movements rely on complex chemical reactions that proceed rapidly when warm and slowly when cold. Imagine what it would be like to live in cold water as the warm-blooded mammals we are. Staying warm would be impossible. Even at 70 degrees F, water pulls heat out of our bodies dramatically faster than does air at a similar temperature. At 40 degrees F water temperature, our bodies couldn't keep up; we would soon die from hypothermia. Mammals that live in water (like whales) or spend a lot of time in it (seals, sea otters, walruses) have special adaptations that include a thick layer of insulating blubber, heavy fur, or both.

Cold-blooded creatures are perfectly fine in cold water. They don't have to heat themselves, which means they don't use a lot of energy except to move around. They don't have to eat much in winter because the cold tamps down their metabolism. Ice anglers can see evidence of this in the fish they catch and clean—the stomachs are often full of food. Prey that digests in a day in summer may take a week in winter.

You may have wondered, if you ice fish, why you catch certain species in winter more so than others. It's because fish react differently to the cold. Walleye and northern pike eat plenty as ice forms, weeds die back, and prey fish become more exposed. Bass and muskies, on the other hand, don't move around much and eat just enough to sustain basic functions. So they're not as easily tempted by a minnow dangling from a tip-up.

When the air temperature hovers in the teens and a wind whips over the snow, we can suffer outdoors from wind chill. But being in the water below the ice would feel a great deal worse. We warm-blooded creatures can be glad about having a heated house or cabin and a bowl of soup to come home to after a walk.

How they breathe down there

As a kid I once asked a wiseacre friend what he had done in swimming class that day. He said, "We learned to breathe underwater." For a long moment I believed him; I imagined puckering my lips down to a tiny pinhole opening and somehow sucking air from the water. That's crazy, of course. But then the question is how fish manage to breathe. The short and easy answer is: fish have gills. A closer look reveals just how remarkable this ability to breathe underwater is.

Here above the surface, the air we breathe contains about 21 percent oxygen. Water contains oxygen in solution, but at nowhere near a comparable percentage. Even very high-quality water, as in a trout stream, contains no more than about 8 parts per million of oxygen— 0.0008 percent. For a fish to get enough oxygen out of such a scant supply borders on the miraculous.

Fishes' gills actually work on the same basic principle as our lungs: exposing a huge number of tiny blood-carrying vessels (capillaries) to the oxygen source, so that the oxygen can migrate in and carbon dioxide can migrate out. Our lungs contain millions of tiny sacs called alveoli, extremely rich in capillaries, where the exchange of gases takes place. Fishes' gills have a structure of rows and columns of specialized cells, called the epithelium, that can absorb the much smaller concentrations of oxygen found in water. In general shape and form, gills look like the fins of a car radiator. Most fish have four gills on each side. There's a main bar-like structure with multiple branches, like a tree, that gradually branch down smaller and smaller, an arrangement that exposes an enormous surface area to the water.

One advantage fish have is that, being cold-blooded, they have lower metabolism than we do and need less oxygen relative to their size. If fish were warm-blooded and needed oxygen for energy to sustain their body temperature, they would not be able to survive on the amount of oxygen the gills can extract from the water. So it's easy to see how facetious my wiseacre friend was when he talked about breathing underwater. If we want to do that, it is easier to buy scuba gear than to grow a set of gills.

Lateral lines: Almost a sixth sense

Fish have the same five senses we do, and one that we don't. It's the ability to pick up vibrations in the water, by way of structures called lateral lines. One could argue that this ability isn't really a separate sense—just a kind of cross between touch and hearing. What's certain is that we have nothing quite like it.

Fish can use their lateral lines to sense the presence of prey or predators even before they can see them, to detect water pressure and thus depth, and to sense the movement of currents and so orient themselves in the water. Scientists believe that in some fish, lateral lines help enable schooling behavior. Lateral lines are common to all fish but are more developed in some species than in others. In many kinds of fish you can see evidence of these structures as a faint line running down the length of each side, from near the gills to the tail. What you can't see is the actual anatomy that makes the lateral lines work.

The lateral line system consists of small receptor patches (neuromasts) in fluid-filled canals on or under the skin. The neuromasts contain hair cells much like those of our inner ears. On these hair cells are smaller hairlike structures called cilia. When the cilia vibrate, the energy of movement is transformed to electrical energy, which travels by way of the nervous system to an area of the brain close to where hearing is processed. The hair cells and cilia are surrounded by a jelly-like structure called a cupula. All this anatomy lies within grooves in the fish's body.

The specific qualities of lateral lines depend on how a given fish species behaves. For example, fish that swim most of the time, as opposed to lying in wait for prey, tend to have lateral lines farther away from the pectoral fins, the pair on the underside just aft of the head. This helps limit the pickup of vibrations caused by the fish's own fin motion.

The lateral lines can be extremely sensitive. Fish in rivers, for example, can detect the vibration of a person walking heavily on the bank near the water's edge. The lateral lines might also pick up the vibration of a baitfish and reveal its location and travel direction so precisely that the fish can turn and grab it even before seeing it. More often, fish use information from their lateral lines along with vision and other senses to locate and identify prey precisely before closing the deal, as it were. So if you want to get close to fish, whether angling or just observing, remember the lateral lines and approach quietly, with as little water disturbance as possible.

Tales in scales

Alive without breath, as cold as death
Never thirsty, ever drinking
All in mail, never clinking

The answer to this riddle from J.R.R. Tolkien's *The Hobbit* is fish. And the "mail" refers to fish scales, structures that can tell a great deal about the fish in your lake. By looking at scales from fish taken during test nettings, scientists can tell how old the fish are, how fast they've grown, whether they've been seriously ill or stressed, and more.

The species of fish in our northern lakes are hatched with all the scales they will ever have. Scales originate from points in the fishes' skin and overlap like shingles in a roof. They grow larger as the fish ages by adding to the outside edge. The scales show growth rings somewhat like those seen in the cross-section of a tree trunk. A difference is that while trees add just one ring per year, a fish scale may gain multiple rings in a year. Still, each year does leave a distinct mark. Because fish are cold-blooded, their growth slows significantly as they spend winter under the ice. A thicker ring, called an annulus, forms during these times.

Biologists can learn a lot by studying these rings. For example, the distances between the annular rings reveal the approximate length of a fish at each age up to the current one. That's because the rings are placed in proportion to the total length of the fish. So, suppose a scale from a 12-inch walleye, three years old, has its first ring one-third of the distance from its focal point to the outer edge. That fish would have been 4 inches long at the end of its first year.

Using scales to tell the age of individual fishes, biologists can learn about the growth rate of a lake's fish population. Because fish grow

Fish Scale
Annual Rings

Eric Roell

more slowly when they reach sexual maturity, scientists can use rings on scales to estimate the age at which fish began to spawn. This helps in setting fishing regulations to make sure most fish can spawn at least once before they are caught and removed.

Fish that swim fast and live in fast-flowing water often have small scales, while fish like carp that live in slow or still water tend to have larger scales. Some fish have smaller scales toward the tail, providing more flexibility there. If a fish loses a scale, such as to injury, it grows back, all at once, so that scale will not show growth rings until subsequent years.

Minnows? Are you sure?

When you see a school of little fish near your pier, you might reflexively think, "Minnows!" But there's a good chance that is both nonspecific and scientifically incorrect. We tend to apply the "minnows" label to any small fish. That's probably because we refer to the baitfish we buy at the tackle shop as minnows, again not precise, but a well-accepted term.

Scientifically speaking, minnows are a family of fish defined by characteristics, not size. Members of the minnow family have one brief dorsal fin with nine or fewer soft rays. They have smooth-feeling scales that may come off when the fish are handled. They do not have true spines in their fins. They have no teeth in the jaw but have rows of toothlike structures on the bony frame that supports the gill tissues. Their teeth are in the throat and help grind food. Most minnows actually are small—the shiners we use for bait are of the minnow family. But the clan also includes carp that can grow very large indeed.

Chances are the schools of fish you see beside your pier are not minnows but the juvenile stage of game fish or panfish. If you can net a few, admittedly not easy, you'll get a clue to what's breeding in your lake. It's amazing how early in life young fish take on the markings of adults. Baby smallmouth bass, for example, have the signature black-tipped tails and red eyes. Largemouth bass have the black stripe along the side, northern pike the oblong oval spots. And so it goes. The young fish seem to mimic adults in temperament too. Little muskies, for example, are hyperaggressive.

FISH BEHAVIOR: HOW THEY ACT AND SURVIVE

Why do fish jump?

You've probably heard the Gershwin song: "Summertime and the livin' is easy / Fish are jumpin' and the cotton is high." On seeing fish jump, you may have wondered: Why do they do that? First we need to qualify the "why" with the reality that fish don't make a conscious decision to jump; they are acting instinctively in response to some stimulus. That coldly logical point aside, it is fairly obvious why certain fish jump. Salmon and steelhead trout, for example, leap out of the water during their spawning migrations to clear obstacles on their way upstream. The obstacle might be a low waterfall, a dam, or a rung on a fish ladder. Asian carp, the invasive species that infests the Illinois River and threatens Lake Michigan, jump in a fright response to boat engine noise.

But then there's seemingly random jumping, like the fish that leaps out of an uncovered aquarium and is found dead the next day on the floor. Or the fish you see jump while you're out in a boat. One reason fish jump is to escape a predator. If you've seen tiny fish leaping out of the water, you can pretty well guess some larger fish is chasing them. But that doesn't account for the jump of, say, a mature largemouth or smallmouth bass.

I've seen it suggested that fish jump to "get a breath of air." This of course is absurd, since fish have gills that are adapted to gather oxygen dissolved in water; they can't breathe air. Some incredible degradation that badly depletes a lake's oxygen might account for jumping, I suppose, but that's surely not an issue on your lake or mine. Another possible explanation is that jumping fish are themselves predators, darting up from below prey fish and in the process shooting clear of the water.

Yet another theory: fish that jump are trying to feed off flying insects that hover near the surface. It's not hard to imagine insect eaters like trout leaping for this reason, but that doesn't seem plausible for lake-dwelling species. Even trout for the most part just slurp floating bugs off the surface. Finally, I've seen suggestions that fish jump for the pure pleasure of it, as an expression of joy. I might stipulate that this explains certain behaviors in higher creatures—otters sliding, monkeys frolicking, dolphins leaping—but I have a hard time buying the idea that fish in our lakes are conscious pleasure-seekers.

So why, then, do they jump? Possibly for almost any of the reasons given above, and possibly for none of them. Maybe in certain cases the answer to why a fish jumps is that there is no why—it just does so some-times. If so, that's fine. We can still enjoy the spectacle.

Who decides where the school goes?

Looking down at a swarm of black-striped fry, moving in unison, a person can't help wondering what holds that school of fish together, why they are schooled in the first place, and which fish decides where the school goes. The first thing to appreciate is that these fish don't decide anything. They don't form the school out of conscious strategic thinking. The behavior is built into their genes; it conveys advantages that promote survival.

For one thing, it's easier for a predator to track down and capture a solitary fish than to eat fish in a school. This seems counterintuitive, since we would think attacking a school would amount to the proverbial shooting fish in a barrel. However, scientists have found that a school confuses predators. A school moving together, the sides of multiple small fish flashing in the sun, can appear to a predator as one large fish, discouraging attack. In addition, the sheer number of fish in a school disorients predators, making it hard for them to zero in on one individual.

Another advantage to schooling is that more eyes watching means greater ability to find food. Schooling also helps fish conserve energy:

in effect they draft on each other, the same principle employed by bi-cycle racers, one closely following another to reduce wind resistance. As for the school's movement, no one fish leads the others. The fish main-tain a fairly precise spacing and, as they swim, they follow their neigh-bors' movements, so that essentially all change course as one. The fish stick together mainly through their vision, relying on markings such as a dot or stripe on the bodies, fins, or tails. They also receive information about neighbors' movements through their vibration-detecting lateral lines.

If you look closely at a school of fry, you may notice individual fish contentedly picking off white specks in the water, likely some form of zooplankton. You can hope the schooling behavior helps those fish grow to catchable, edible size. Time will tell.

Bass beds, no strikes

If your lake contains bass, you'll start to spot circular spawning beds in the shallows around the end of May or start of June. If you're an angler, you may cast to them with limited results, for some days anyway. It's said that catching smallmouths or largemouths from spawning beds is so easy as to be unsporting. That is true, but only at the right time in the spawning cycle.

When the water temperature just edges into the upper 50s or low 60s F, that's the trigger point for spawning season. The males move into the shallows and use their tail fin to sweep away sand and expose gravel to create a spawning bed, slightly bowl-shaped and about 3 feet in diameter. The males then wait for females to arrive. At this stage, the male bass are not the least bit interested in food. They spend the time after ice-out, the period known as pre-spawn, feeding heavily, but when spawning time arrives, they basically quit eating. A friend once explained this food abstinence as clearly as anyone could: Throwing bait in front of a spawn-minded bass "is like someone offering you a hamburger while you're in the throes of passion."

Once the male finds a mate and the eggs are laid and fertilized, the male's job is to guard the nest, and later the fry, for a couple of weeks. Here is where the fishing gets exciting. It's not that the bass have started eating again. It's that they aggressively attack and remove anything that encroaches on the nest where the eggs lie on the bottom incubating. That includes a lure dragged across the bed or a leech or nightcrawler dangled above it.

There are some who say anglers should leave spawning bass alone. Others say that if released unharmed, the bass return almost immediately to the nest and no harm is done. For my part, I mostly prefer to leave them alone and let the reproductive process play out. Now as I ponder the spawning ritual, I can't help asking: How many women wish their men would behave more like smallmouths: males who make the bed and take care of the kids?

The making of a magnet

The shoreline here on Birch Lake is relatively barren of fish-attracting fallen timber, but not long ago I got to watch one wooden specimen make the transformation from shade provider to fish haven. Just down the shore from our pier stood a tall white pine, its roots right at the waterline, its imposing trunk angled over the water at about 30 degrees from the vertical. It seemed to defy the tug of gravity, and I hesitated to paddle beneath it in a canoe. I wondered if it ever would tip into the water.

Well, before long, I got the answer. One midsummer day, I saw the tree had tipped to about 45 degrees, and I noticed a large, lengthwise crack at the base of the trunk. After a few weeks, the old pine came down, but not with a spectacular splash. It eased down, like a staccato second hand on a watch, tick, tick, tick. During a couple of quiet evenings, sitting on the screen porch, I could hear the periodic cracking noises as the tree kept ticking down. Then one morning the tree lay in the water, extending about 70 feet out from shore.

It was sad to see a venerable pine go down, but the plus side is that the tree now lies in what already was a fair walleye hole, just off the edge of a bed of bulrushes, at a U-shaped dropoff that anglers like to call an inside turn. Snorkeling around the tree soon after it fell, I saw young-of-the-year smallmouth bass darting amid the twigs and still-green needles. Once a perching place for bald eagles, the old pine had become a fish magnet.

How do you like these odds?

If in early summer you see large schools of fry in the shallows, you might assume your lake will soon be chock full of fish for catching. The reality is far different. The odds of survival for these fry are exceedingly long.

One scientific study used DNA tracking to estimate the success of spawning smallmouth bass on a lake in Ontario. The study found that only 27.7 percent of male bass had at least one offspring survive to the fall young-of-the-year stage. Just 5.4 percent of all the spawning males produced 54.7 percent of the total number of the fall young-of-the-year.

To look at it another way, consider that female smallmouth bass deposit two thousand to ten thousand eggs on a spring spawning bed. Even under the best conditions, most eggs don't survive. They're vulnerable to changes in water temperature and oxygen levels, flooding or sedimentation, disease, and predation from panfish and crayfish.

When the eggs hatch, the larval fish live off a yolk sac attached to their bodies. Once the yolk is fully absorbed, the fry, about 1 inch long, rise from the bed and start eating on their own. They survive on tiny crustaceans until they are big enough to eat aquatic insects, then larger crustaceans and fry of other fish that spawned later. As the fish grow, they face some of the same threats as the eggs, in addition to which predators feast on them.

When they're small, they get attacked by bluegills, perch, and sunfish. As they grow, they become prey for walleyes, northern pike, and muskies. Other enemies, again depending on the fishes' size, include kingfishers, loons and herons, mink, frogs, and some snakes. The result is that only a tiny fraction of the eggs laid in a spawning bed, and only a few of the fry you see near your pier, ever become adults. Nature can be a cruel mother indeed.

How fast do fish grow?

In the brutal winter of 2013-14, one of my favorite northern pike lakes froze out. When a friend and I fished it the following summer, we caught just one pike, and a fair description of it would be anorexic—a pitiful sight. I wondered then, if nature were left to its own devices, unaided by stocked fish from hatcheries, how long it would take to catch keeper-sized pike on that lake again. And for that matter, suitable perch, bluegills, smallmouth bass, crappies, and walleyes, which also inhabited the lake.

Growth rates vary with a number of factors, chiefly food abundance and water temperature. What you'll see here are averages based on information from departments of natural resources in Upper Midwest states. Let's start with northern pike. These guys grow fast. Males become mature in a year or two, females in two or three. Pike typically reach 10 inches long after one year and 18 inches after two. That 24-inch pike you catch is probably just three years old. Generally speaking, females of any species are somewhat larger than males at a given age.

Yellow perch, a favorite pike forage, grow fairly fast in length for their first few years and after that mainly put on weight. They'll be 3 to 5 inches long after one year, 5 to 9 inches after three, and 7 to 10 inches after five. After seven years they'll have grown to only 10 to 11 inches but will have gone from 0.4 to 0.7 pounds.

Bluegills? That 5-inch borderline keeper you catch might be as old as three years. The 8-inch "bulls" many of us yearn for might be seven to nine years old. As for black crappies, they're 3 inches long after a year, 7 inches after three, and nearly 10 inches after five. A prize 13-inch slab is likely ten years old.

As for largemouth and smallmouth bass, a 14-incher is probably five years old. If you catch one 18 inches or larger, it's most likely in its ninth or tenth year. There's a practical lesson anglers can take from this. When deciding which legal fish to keep, or how many, we should think about how long those fish will take nature to replace. Then keep or release as we consider prudent.

When fish can't grow

Years ago a friend told me about a small lake that was great for bass fishing. Another friend and I tried it on an August Friday, just after sunset. Sure enough, my second cast of a floating lure brought an explosive strike, and a hard rush, but then . . . nothing. The fish just gave up and let me reel it back to the boat. All the bass we hooked that evening behaved the same way. They were maybe 12 inches long, but quite skinny and unable to fight.

Anglers would say the bass in that lake were stunted. That would be true, and the same could be said for abundant small bluegills in other lakes. However, the stunted growth likely says less about the fish themselves than about the lake's environment or the prey-predator balance. Fish we consider stunted are likely not inherently unhealthy. It's something about the lake that keeps them from thriving. Put those same fish in a lake with more suitable conditions and they would grow to normal size.

Fisheries biologists say that some lakes have stunted panfish because of overfishing: Anglers keep the bigger fish and toss back the little ones;

pretty soon there are few bigger ones left. In this case the fish are not really stunted—the growth rate may be just fine but the harvest is too heavy. Often, however, the cause of small fish is very slow growth. The reason may be overabundance (too many months for the food supply), too much cover leading to inefficient predation, too few predators, or a combination of these factors.

Cures for stunted fish populations aren't simple. For bluegills, remedies that have been tried and mostly have failed include netting and removing large numbers of smaller fish, clearing out protective vegetation to expose the small fish to predation, and poisoning that kills smaller bluegills but doesn't harm the bigger fish. One promising approach to fix bluegill stunting, according to biologists, is to use size-limit regulations to foster a population of largemouth bass in the 12- to 16-inch size range. Bass that size feed heavily on young bluegills.

Fish kills: Not just a winter event

One event anglers dread is a winter kill on a favorite lake. While mass fish die-offs like those on shallow lakes are traumatic, the fact is that fish die all year long, just not in big numbers. And winter kill is not the only kind of seasonal fish kill. Most fish die from natural causes, like any creature. In fact, fish live relatively short lives and can die from injury, predation, starvation, disease, parasites, severe weather, old age, and even lightning strikes. Even among game fish too big to be eaten by predators, about half typically die in a given year.

As for larger-scale fish kills, they can occur in spring, summer, and winter. Some are dramatic: fish appear floating belly-up or washed up against shorelines. Others are less obvious: the fish just sink to the bottom and decay or are eaten by turtles, crayfish, and other scavengers. Summer kills, not especially common in northern lakes, can happen after long spells of very hot weather, mostly in shallow, nutrient-rich lakes with abundant algae and weeds. The plants and algae use up oxygen overnight, leaving low dissolved oxygen in the water as the new day dawns.

Conditions get worse if the weather stays cloudy and calm, as there is less restoration of oxygen from the plants' photosynthesis and from the mixing action of waves. A dissolved oxygen level of less than 2 parts per million endangers most fish, and they will die if the oxygen is not recharged. Sometimes the fish, stressed by low oxygen, become susceptible to deadly infection by viruses or bacteria.

Winter kill is much more common in northern latitudes. Again, shallow and weedy lakes are the most vulnerable. Although the dead fish are discovered in spring after the ice leaves, the fish actually die in late winter, from suffocation. Spring kills can happen as the water warms up after ice-out. Fish may come through the winter weakened from eating less than in the open-water months. They become stressed as their metabolism increases and they expend substantial energy in spawning. Spring kills generally don't affect large numbers of fish, and even winter kills usually don't wipe out all the fish in a lake. Enough remain to reproduce and restore the population over several years.

ANGLING INTERLUDES

The $100,000 bass

Every lake home or cottage owner has a story of how it all began. We live on Birch Lake because we vacationed here with our kids for over twenty years and came to love it, but also because of one particular largemouth bass and, I am convinced, cosmic forces. It was a warm and gray Friday in August 2009, the last day of our week's stay at a rented cottage. Fishing conditions being ideal, I climbed into the boat and motored straight across the lake. There on shore, partly hidden by tall rushes, stood a For Sale sign. Over the years, my wife, Noelle, and I had window-shopped for properties on the lake; steadily rising prices kept our dreams of a Northwoods place elusive. We had looked somewhat seriously at a lot with a small, run-down red cabin on it, but the asking price seemed high.

Now, here I was, in my boat looking at a vacant lot, heavily wooded with red oak, hemlock, and a few majestic white pines. The top of its steep slope would afford a great lake view. It had been for sale for a couple of years, and with the recession the price had dropped from "Are you kidding?" to "Possible." On the last day of vacation, carrying years of wishing and hoping and in pain at the prospect of leaving for home the next day, I said to myself, "All right. If I catch a fish in front of this For Sale sign, we are destined to buy this property."

I launched a black-and-gold jointed Rapala floating plug that landed just where I aimed it, up against the edge of the rushes. One twitch, two, and . . . *BAM!* It turned out to be a near-trophy largemouth bass, extremely well fed, beautifully colored, black lateral stripe on deep green.

I slipped the bass back into the water, headed back to the cabin, and said to Noelle, "Guess what?" We agreed that destiny had called. A month later, in the full-color splendor of late September, we walked the site with a real estate agent. Three months later we signed closing papers at a bank office in Minocqua.

In April 2011 we held a ground-breaking ceremony, in a snowstorm, with our two kids and their significant others. By that September, we had a cozy little lake home. So here we are, full-time lake dwellers. And somewhere in Birch Lake, there swims a very expensive fish.

The zen of the jig

If walleyes inhabit your lake, you are privileged. Walleyes are the essence of angling in the northern latitudes. We have plenty here at Birch Lake. They run a little small; on a good day we need to sort through several to find a couple that meet the 15-inch legal size limit (and I don't keep many anyway). I sometimes wonder: What's the attraction? Fishermen love a fight, and walleyes when hooked tend to come along quite peacefully. They'll hold deep for a while but then give in.

Walleyes may be the best-tasting freshwater fish alive, but there's more to it than that—mainly the finesse it takes to catch them. You set up over a favorite rock bar. You know they are down there, those greenish-gold spooks with eyes like precision-ground lenses. I'm not sure how they manage to be so light on the take. They seem to lift a jig and bait from the rocks with the touch of a pickpocket, and they will pick you clean if you're not fully attuned.

You move the jig a small twitch at a time. Through a sensitive graphite rod you can feel every bump on a stone and the tug of every weed. You wait for a sensation that is just a bit different, a subtle *tik-tik*, or sometimes just a bit of extra weight that ever so slightly pulls back. A snap of the wrist, and you're fast to a fish. Even when fishing with slip bobbers, the walleye bite is distinctive. The bobber will twitch, pause, and then go down. What I imagine happens is that the fish noses up to the bait, stops, inhales it with a jet of water through the gills (the twitch), pauses for just an instant, and then moves on.

Jigging, though, is the way to experience walleyes. It forces you to be fully in the moment. Where walleyes are concerned, it is impossible to jig successfully without being nearly in the state of zen. As for tangible rewards, a meal of fresh-caught walleyes may not qualify as a sacred feast, but it's close.

Playing the float

There are various non-powered ways to enjoy your lake: canoe, rowboat, kayak, paddleboat, paddleboard, air mattress. There's one more you've likely never tried: the float tube. I had read about float tubes in books about trout fishing. I had never actually seen one until I ran across a used model for sale. New tubes can retail for $150, $200, $300 or more. This one was priced at $25. Sold.

Good intentions being what they are, I let the thing sit around my basement unused for three years. Then I finally tried it out. It's like the upper half of a big inflatable armchair. You step through a couple of holes in a webbing "seat," put on flippers, and wade out into the lake until buoyancy takes over. You move around, backward, by doing a basic straight-legged backstroke kick. A float tube would be great for just lounging in the water and enjoying a beverage, but these contrivances are made for fishing. Mine has a couple of zippered compartments on each arm, great for storing containers of bait, a small tackle box of basic items, a needle-nose pliers, and other gadgets.

I soon found that with vigorous kicking of my flippered feet I could cover reasonable distances to get within casting range of attractive features like downed trees and clusters of lily pads. I wouldn't try crossing a lake in the tube, but I can range a decent distance from the pier and not worry about running out of steam on the return trip, even if I have to kick upwind. There's a kind of elemental pleasure in hooking and reeling in a fish while suspended waist-deep in the fishes' environment. It's more satisfying in that respect than wading. You play a fish not just in front of you but beside, behind, and underneath. It doesn't take a very big fish to tow the float tube around a bit.

There is of course the risk of dropping the rod into the water and losing it forever during basic tasks like baiting up, squeezing on a new split shot, or working a hook out of a bluegill's jaw. You've got to be careful how you lay the rod down to free your hands. There's also the matter of wearing a life vest. I wouldn't go tubing without one, especially after the time my tube sprang a leak. I did have a vest on, but still I beat it for the shallows as the air hissed out. Once safely in knee-deep water, I found the main inflation valve had opened. It was an easy fix and I was soon back in business.

I still keep the tube in the garage, but during summer in an inflated condition, not folded and stuffed in a gym bag. It's great fun on my own lake and for exploring those little woodsy ponds nearby, not easily accessible even by canoe. It's a new twist on silent angling.

In the middle of nowhere

I often hear lake home and cottage owners say, "My lake is no good for fishing." Maybe that's true, or maybe some people who say that are just using the old angler's trick of keeping their mouths shut about good spots. Or maybe they just haven't discovered the magic of something called the midlake hump.

Serious fishermen know about humps. They're also known as bars, reefs, or sunken islands. Call them what you will, they hold fish. If you see a boat parked out on your lake, away from shore, seemingly in the middle of nowhere, chances are that angler is working a hump. Novice fishermen, and even longtimers who never got serious, tend to fish along the shore, in the lily pads, around piers, or in fallen timber, a tactic disparagingly called bank beating. That can work, but most times you'll find more and bigger fish around deeper-water features. I am nobody's fishing expert, but I do know about humps. The easiest way find them is to buy a topographic map of your lake. If no such map exists, you'll have to go exploring.

You can locate humps by prowling around slowly with your depth finder (fish locator) running. Before I owned one of those, I found humps by rowing or motoring slowly with the anchor hanging down. When it hit a hump, I knew. It's not difficult to work a hump. You fish pretty much the way you would in a nearshore area. If the hump is shallow, you can cast spinners. If it's a little deeper, throw deep-running crankbaits. In either case, you can hop jigs along the bottom, or cast slip-bobber rigs.

Humps can be so productive that once you know about them, you may give up bank beating for good, except in the times of year, like early season, when fish are known to inhabit the shallows, and if pursuing bass that like to lurk in near-shore cover. If you've been mostly a bank beater on your lake, try finding a midlake hump. You might discover a great place to fish with your kids, and you might never again have cause to say, "My lake is no good for fishing."

Opportunity fishing

One early May Saturday on the fishing season opener, I stayed off the water. The day dawned cold with a stiff northeast wind and a crystal sky, about the worst conditions for walleye fishing one can imagine. So I didn't go out. No law says I had to. I'm no more obligated to fish on opening day than to drink to excess on New Year's Eve. One benefit of lakefront life is being able to fish—or not—as I choose. I remember opening days coming north with friends and encountering miserable weather. We fished anyhow; we had planned the trip for months, booked a cabin, driven four hours, spent money. Was it fun on the water? Not exactly. What fish we caught we paid for dearly in discomfort, and sometimes we caught nothing. Evenings around a card table or campfire made the trip worthwhile, but we could have had that without the fishing.

The thought process is different when the lake waits just down the hill. If opening weekend is dreadful, why bother? There's next weekend, and then any weekday evening after work. Nothing is lost except a couple of days on the calendar, and there are lots of those to come, week upon week, month after month. I like to call this approach opportunity fishing—a privilege elusive to those who need to travel to a favorite lake and make advance arrangements. The late outdoor writer Gordon MacQuarrie explored this concept in a classic story. Though he worked in Milwaukee, his friend Gus lived in the north and would call him when the flights of ducks were moving over the marshes. Every hunter or angler, MacQuarrie said, needs a Gus.

Now, I suppose I'm Gus to my brothers and friends who still live to the south. They can't always come if I call to say the walleyes are biting, but I'm ready to go almost anytime the conditions are right. There's also the advantage going out only for the hour or so of prime time. Here on Birch Lake that typically starts when pines across the lake just brush the setting sun and ends when it gets too dark to see a slip bobber.

So, in that recent year I allowed opening day to slip by. Well, all right, not completely. I did go out for an hour or so in the evening and hopped a jig and minnow over some rocky areas where I know that walleyes prowl. The wind blew the boat around, I had not one strike, my hands got cold and a little stiff, so I packed it in just as the sun went down. I fished on the opener. I might as well not have bothered. And that just helps confirm the luxury of opportunity fishing.

Bobber down

For those not interested in fishing, watching a bobber may be the world's most boring way to spend time. For those who do fish, it's exciting or incredibly relaxing, and sometimes both at once. We associate bobber fishing with kids, but I didn't truly discover it until adulthood. When I was little, I thought fishing with a bobber meant using a minnow; it was how my dad fished, and I seldom saw him catch anything.

Using a bobber was, well, boring, compared to casting a piece of worm on a hook and getting regular bites from panfish. Besides, it was hard to cast one of those red-and-white spring-clasp bobbers with 6 feet of line dangling. (If slip bobbers existed back in the 1960s, I wasn't aware of them.) I learned the virtue of bobbers in my midtwenties on a weekend fishing trip to Lac Vieux Desert on the Wisconsin-Michigan border. One evening two friends and I caught half a dozen walleyes by suspending shiners on bobbers just over the tops of cabbage weeds. Since then my tacklebox has always carried a supply of floats in different sizes and colors.

I fish with various methods, using lures and live bait as the situation requires, but I cherish the times and places where slip bobbering is the ticket. Once in a while the bite is so fast and furious that the bobber spends little time afloat. More often, between bites, there are long intervals of meditative attention, the modern term for which is mindfulness. If it's really impossible to fish and worry at the same time, that is especially true when watching a bobber.

A bobber has ways of signaling strikes other than just submerging. Sometimes it skitters across the surface—a fish is swimming away with the bait. Now and then it just tips over on its side, a sign that a fish has taken the bait and moved up, not down, in the water column. Other times the bobber just dips a little lower in the water as if another split shot has been added to the line; here the fish is mouthing the bait and finning in place. In any case, I pick up the rod often not knowing for sure what species of fish, and how big, waits at the end of the line. I reel down until I feel slight pressure, then sock the hook home.

Fast action or slow, bobber watching is solace for the soul. Picture this. You've just arrived at your lake place for a week's vacation. You get in a boat toward sunset, row out to a favorite spot, deploy a leech on a bobber, sit back, and watch. You know there are more sophisticated, more adult ways of fishing, and yet here you are, fishing like a child. You see the float on the still surface twitch and then disappear. It's the start of a bountiful evening. And best of all, late at night, you can drift into sleep to a loop video in your head of just one thing: a bobber going down. It definitely beats counting sheep.

Cherishing the moments

Before I lived here in the north, my fishing season typically ended around Labor Day; maybe one more outing in mid-September. In fall, the fish were no longer in their midseason haunts and thus were harder to find.

Besides, the weather was chancy: pick a weekend for a trip and the risk of hitting a cold front or rainstorm ran high. Now, though, I have come to prize September and October angling.

Birch Lake is largely deserted after many seasonal folks close up their cabins around Labor Day. There's no competition for my favorite spots. In fact, most often, especially on weeknights, my boat is the only one on the lake. After nearly thirty years here, as an annual visitor and more recently a resident, I know the water well. It doesn't matter what the guide on the radio says about where the walleyes and smallmouths can be found on the area lakes. I know where they are on Birch Lake, and that's really all I need.

But it isn't just about the fishing. Out on the water, the autumn air is bracing, especially as the sun touches the treetops and the evening cools. The water temperature falls day by day; the algae die back and the water clears. The evenings are incredibly quiet. There's no roar of boat engines, just the occasional call of a crow or the scream of an eagle; mostly just pure, profound stillness that seems to clear the head of the internal echoes from the noise of daily life. And there's nothing quite like blue water to set off the fall leaves; at peak time it's as if some deity took hold of a cosmic knob and dialed the color up to maximum.

Typically, by mid-October, I've fished for the last time of the year, the pier mostly disassembled, the board sections and frame pieces stacked on shore, the boat ready to be pulled out and stored. I've made the most of autumn on the lake. As winter sets in I won't have to regret precious times missed. I've said my good-bye; I'm ready for the ice.

SPECIES WE TREASURE

Walleyes

UNDER COVER OF DARKNESS

Soon after the ice leaves our lakes, there plays out a ritual of biological and cultural importance: walleyes moving into the shallows at night to spawn. When it happens is mainly a question of water temperature: the sweet spot is 45 degrees F, give or take. Where it happens is a question of lake-bottom characteristics: walleyes prefer rocky or gravelly bottom in water roughly 2 to 6 feet deep.

To observe the ritual, you need to know where your lake's spawning areas are, get there on the right evening, and have a good flashlight—the beam catches the fishes' highly reflective eyes. It's fitting that walleyes, being night prowlers by nature, breed under the moon and stars. The action starts around dark and lasts until about midnight. It is at the same time a mass migration and a fleeting event. While bass spawn in solitary fashion, fanning out individual nests in sandy and gravelly bottom, walleyes make it a convention.

The males move into the shallow water first, staging for the ritual. A few days later the females arrive, generally larger fish. During active spawning, you'll observe a female spawning with one or two males. The spawning group swims in circles, the fish sometimes pushing and bumping each other. Then they stop; the female turns on her side and drops as many as half a million eggs. The males shed their milt over them. The eggs sink to the bottom and into crevices in rocks and gravel that afford some protection against predators. Still, only a small percentage of the eggs hatch, and a still smaller share of the fry ultimately survive. Preservation of a species in fish is a numbers game. The females usually discharge most or all of their eggs in one night of spawning and then leave the area. The males stay behind for a few days, but not to guard the eggs in the manner of bass. Walleyes are broadcast spawners—fry that hatch are on their own.

Even if we don't get to watch the spawning, because we don't have access to the sites or because our timing is wrong, there's hope in knowing that it's happening. Walleye spawning is a good sign of health in our lakes, a forerunner of quiet evenings in a boat, rod in hand, and a reminder of the everyday miracle of reproduction. With the songs of frogs from the woodland ponds and the call of the season's first loon, it's an essential rite of springtime.

WALLEYE VISION

The defining characteristic of walleyes is the eyes, like clear marbles. The eyes are the key to understanding walleyes' behavior and their advantages in the game of survival. If you've ever aimed a flashlight at walleyes in shallow water, such as in spring while they're spawning, you've seen those eyes shine. You may also have seen this glow in flash pictures of walleyes taken at night. That's caused by a layer of reflective crystals behind the retina. Called the *tapetum lucidum*, it gives walleyes the excellent vision that makes them deadly predators at night or in murky waters.

Walleyes feed largely on perch and other smaller fish that don't see nearly as well in low light. In an important sense they are stealth predators, approaching almost (to the prey) invisibly. Faint light entering the lenses of the walleye's eyes passes over the rods and cones, basic structures that enable vision in many creatures, including people. The light then hits the *tapetum lucidum* and is reflected so that it passes over the rods and cones again. This extreme sensitivity to light explains why walleyes during daylight take to deep water or hide out in the shade of weed beds. On many lakes, even some where the water isn't especially clear, it's hard to find walleyes when the sun is out. The best times are cloudy days and around dawn and dusk.

Despite the quality of their vision, walleyes don't see colors as well as other fish, such as bass and northern pike. Researchers believe they see all colors in some shade of red or green. In low light, they don't see color at all, just shapes and shades of black and white. Studies also show that walleyes respond to changes in light intensity: rapid changes can trigger feeding spells. They're also active when the wind kicks up waves. The rough surface diffuses light; the wave action can also disorient baitfish, making them easier to capture. It's no accident that anglers say they have more success when there is a "walleye chop" on the water.

Walleyes also have very sensitive lateral lines that pick up vibrations. This helps them detect prey fish swimming erratically. They also have excellent hearing—a noise, such as from someone banging around in a boat above, can scare them into fleeing. So walleyes are well adapted to live in our lakes, and to frustrate anglers a goodly share of the time.

Bucketmouth and bronzeback

It's really not hard at all to distinguish smallmouth bass from large-mouths. The most common advice cites differences in the mouths. In largemouths the hinge of the jaw lies aft of the eye; in smallmouths it's even with the middle of the eye. The difference in mouths gets more pronounced as the fish grow bigger. Not for nothing are largemouths nicknamed "bucketmouth." Open the maw of a five- or six-pounder and you could pretty well fit your fist inside. This mouth is great for engulfing large prey. A smallie's mouth, more in proportion to the fish's size, has its own benefit: it's great for surgical strikes against crayfish, a favored food, hiding among rocks.

Other physical differences are also quite apparent—there's no excuse for failing to tell these species apart. Largemouths are dark green and have an irregular black stripe running the length of each side; small-mouths are greenish-bronze (the nickname "bronzeback" fits) and have vertical bars on their sides. Smallmouths are also distinctive for their bright-red eyes. The most compelling way to distinguish smallmouths, though, is to get one on the line. Their fight is tenacious and powerful far out of proportion to their size. Largemouths can put up a battle too, and like their cousins they'll leap out of the water with abandon, wildly thrashing in the air. Still, pound for pound, there's no doubt who wins in this comparison.

Some anglers greatly prefer smallmouths for that reason, and some for yet another: they are a little bit like trout. If you see bass on a weedy lake or in the warmth of a calm river backwater, those are probably largemouths. Smallies favor lakes with rocky and gravelly bottoms. They like their water a little cleaner; being fairly intolerant of pollution, their presence in a lake indicates water quality that's reasonably good. They also like things a few degrees cooler and often inhabit fast-flowing streams (though not as cold as those favored by trout).

There's almost a bit of class distinction between the two fishes. If largemouth bass are beer and bluejeans, smallmouths are brandy and khakis. If largemouths are tavern brawlers, smallmouths are fine-tuned, highly trained fighters.

Northern pike: The people's fish

Northern pike often don't get the respect they deserve. While some anglers consider pike inferior to fish like walleyes and bass, they are a worthy species. Northerns go by a variety of nicknames: gators, snakes, water wolves, and (in Canada) jackfish. Perhaps the most appropriate label is "the people's fish." For one thing, northerns are abundant, and for another, catching them doesn't require any special angling skill. Put a lively minnow in front of them, or drag a shiny lure past them, and they will strike with abandon.

If there were a drag race for Wisconsin fish, northerns would be among the top contenders. They accelerate rapidly from a standstill to a top speed approaching 10 miles per hour, quite a good clip in a water environment. When they strike prey they are quite deadly. The lower jaw is ringed with very sharp conical teeth; short and sharp teeth also line the roof of the mouth and the tongue. A smaller fish that gets caught in those teeth stands no chance.

As they grow, northern pike come to prefer fish with more cylindrical shapes, like suckers and chubs, as opposed to deeper-bodied panfish with spiny dorsal fins that make them harder to swallow. Pike feed voraciously and will keep eating even when their stomach is full; they have been known to choke to death on large prey they catch but can't swallow.

Northerns begin to stage for spawning while ice still hangs around on many lakes, seeking shallow, marshy habitats or the grassy edges of lakes. In late-thaw years, pike may spawn under the ice. This early reproduction confers a competitive advantage: pike fry can grow large enough to eat the newly hatched fry of fish that spawn later.

Many anglers and their families shy away from eating pike because they wear a heavy coating of slime and have numerous, slender, forked bones (Y-bones) in the flesh of the back. Actually, pike are delicious, either fried or baked whole in the oven, stuffed with bread, corn, and onions, and wrapped in foil. For those who prefer filets, there are methods for removing the Y-bones while wasting very little meat. Learn to prepare pike properly and you'll gain more appreciation for their presence in your lake.

Musky: King of the lake

Imagine spending an afternoon on your lake, peacefully kicking along with flippers and snorkel, hovering over a tangle of logs and branches, hoping to see walleyes lurking there, when suddenly you're staring into the toothy jaws of a musky, just a few feet ahead. It's a bit unnerving, but only for an instant, as you realize there's nothing to fear. At your approach the musky calmly turns and swims off. You follow for a while, moving slowly and quietly, until the fish has had it with the game and with swift strokes of its tail darts for the deep.

Chances are you just chased the musky out of a favorite "restaurant." Muskies are the ultimate ambush feeders, lurking dead-still under cover of wood or weeds and then darting out at blazing speed when a prey fish of suitable size happens by. They grab and hold it with sharp canine teeth and then swallow it head first, sometimes retreating to their hiding place before doing so. Besides fish, they'll eat ducklings, muskrats, and frogs. In general, the bigger the musky, the larger the prey. Muskies are mainly sight feeders; they have difficulty hunting at night or in murky, cloudy water.

Muskies themselves are quite vulnerable, as all fish are, during their early life stages from larva to fry to juvenile fish. They're prey to bigger fish, even to their own kind occasionally. But a musky that runs the gauntlet for a few years to become an adult is the king of the lake, without natural enemies, other than anglers hurling heavy lures. Effective as they are as predators, muskies are not as aggressive as northern pike. While northerns will quickly attack almost any prey that enters their field of vision, muskies can be somewhat reticent, as anglers can attest. They'll often follow a few feet behind a lure and then at the last instant turn away.

Muskies come in color patterns that vary with the lakes where they live. The three most common patterns are clear, with only faint vertical markings on their flanks; spotted, with dark more or less circular spots on a lighter background (as opposed to the white, oblong spots on northern pike); and barred, with dark, fairly wide vertical markings. Tiger muskies, actually a musky/northern pike hybrid, have another distinctive coloration with narrow, vertical dark stripes on a lighter background. If muskies live in your lake, there likely will be just one adult for every two to three acres of water. That's why, whether fishing or snorkeling, it's a treat to see a big one.

Crappies: Angels of the lakes

Among the joys of lake life is watching the water world respond to warming conditions in spring. After three straight days with temperatures around 80 degrees F, the fishing guide on the radio said crappies would be moving into the shallows on our lakes.

Crappies, specifically black crappies, are our biggest panfish. Some reach 15 inches long (I caught one that size as a kid). They are also quite beautiful. If you have observed a crappie in an aquarium, you know that with their speckled markings, graceful shape, and large dorsal and anal fins, they are the closest thing in appearance to angel fish that you'll find in freshwater. The fins, though actually quite spiny, appear delicate when crappies are suspended in the water. The mouth is tender; when hooked, crappies need to be treated with care lest the hook tear free. Not for nothing are they nicknamed "papermouth."

They are also delicate in the way they approach and take prey. Instead of chasing and slashing at it in the manner of a northern pike, they come up behind, beside, or beneath it, watch for a moment, and then slurp it in. Their flesh is delicate table fare. Many anglers rank the tender white meat right alongside perch and walleyes for appeal.

One thing you can count on where crappies are concerned is that when they come into shallow water to spawn, you'll find them in the wood. Next to sunken logs, over piles of submerged brush, beside wooden pilings—that's where they'll be. They also strongly favor bulrush beds with a firm, sandy bottom. The crappie bite is a welcome spring ritual, just another of those seasonal certainties that mark the rhythm of lake life.

Bullheads: Whiskered ones

Though chances are you have never caught a bullhead in your lake, bullheads probably live in it. These olive-backed, yellow-bellied, bottom-dwelling fish can be found in most northern lakes. We tend to associate them with muddy rivers and ponds, and they do thrive in such places, being quite tolerant of low dissolved oxygen in the water. They are often among the survivors when lakes experience winter kill. None of this means they can't live where the water is clean. They can, and they do.

Wisconsin has yellow, black, and brown bullheads. I grew up catching bullheads from the East Twin River in my hometown of Two Rivers. I remember them for their abundance but also for the sharp spines, one each on the dorsal fin and the two pectoral fins. As kids we had to learn how to hold a very slippery bullhead so we could remove a hook from its jaw without getting poked. We believed then—and it turns out to be true—that in addition to a puncture wound, the spines deliver a dose of mild toxin that makes the injury hurt all the more. The smaller and younger the bullhead, it seemed, the sharper the spines.

These spines are a defense against more than fishermen. When threatened, bullheads extend the spines straight out. That makes them tough for a predator to swallow, and a stab to the mouth with poisoned-laced spine adds to the deterrent. Like catfish, which are close relatives, bullheads have barbels—whisker-like structures—under the chin that help them find food. As the fish feed, the barbels touch the lake or river bottom; taste buds on the barbels detect the groceries. Bullheads eat almost anything, from living or dead plant material to aquatic insects, snails, crayfish, fish eggs, and other fish.

Bullheads spawn from late spring into early summer. If you've ever seen in your lake's shallows a round, dense school of little black fish, those are likely bullhead fry. For a time the parents actually guard the school, the male acting like a sheepdog, herding stragglers back into the group. The young remain in their schools for a long spell even after mom and dad leave.

My memory tells me bullheads are good to eat, especially when caught from cold, clean water. Mom cooked up quite a few East Twin bullheads over the years. They're a chore to clean, though, since they have to be skinned (they have no scales). If you want to try catching bullheads on your lake, your best bet is to fish at night and soak worms on the bottom. Bullheads feed under the moon and stars.

Sunfish: Freshwater jewels

You've probably seen neon tetras, bleeding hearts, clownfish, and yellow tangs in aquariums. In all likelihood a fish just as beautiful, or more so, swims in your lake. For my money, no fish is prettier than the sunfish, officially the pumpkinseed. I sometimes question why they are named for something as nondescript as those off-white ovals pulled from Halloween jack-o-lanterns. The name "sunfish" belongs to a family of fishes that include the bluegill. Still, what the experts call pumpkinseeds are always sunfish to me.

Bluegills are pretty fish in their own right, but they look pale next to sunfish: Those eyes with bright-red irises around the dark pupils. The wavy lines that radiate from the mouth across the gills, in a color like aquamarine charged by blacklight. The deep black gill spot with the accent of brilliant red. The subtle pattern of blue-and-emerald vertical bars across the golden body. And then that bright yellow-gold belly. Years ago I kept a couple of sunfish in an aquarium. They looked great amid the green artificial weeds, lit by the fluorescent lamp, and our toddler daughter loved watching them.

Fortunately for us all, sunfish are extremely common. They're generally not as abundant as their bluegill cousins, but you can find them in almost any lake. They live in the shallower water and in the weed beds, eating mostly insects and their larval forms. When fishing with kids, nothing will bring more cries of delight than one of these jewels, popped from the lake, multiple colors glistening in an evening sun.

Ciscoes: Seldom seen, always important

Sometimes it's the people behind the scenes who make a business go— the shop foreman, the office manager, the administrative assistant, and others who may never see the limelight. It can be like that in your lake too: creatures you never or only rarely see have a lot to do with keeping the ecosystem healthy. In some northern lakes, those unsung heroes include cisco, a fairly small, silvery species similar to whitefish.

Ciscoes, also called lake herring and tulibee, are common in Lakes Michigan and Superior and are also found in the deeper, colder inland lakes in northern latitudes. Where they exist, they are an essential component of the food web. Few ciscoes mean fewer (or smaller) game fish like walleyes, northern pike, and muskies. In fact, ciscoes are a nearly ideal prey fish by virtue of their size, typically 10 to 14 inches, and their fatty meat. I once heard a fisheries biologist refer to them as "swimming sticks of butter." A study in Ontario showed that walleye grew more efficiently and had to eat less often, thus expending less energy, in lakes with abundant cisco. In turn, ciscoes are a target for anglers: they feed heavily under the ice and are pursued by ice fishermen, who use tiny jigs tipped with waxworms or wigglers.

Ciscoes require cold water and relatively high dissolved oxygen. Even small changes in these parameters can cause die-offs. A decline in the cisco population can be a sign that a lake's water quality is deteriorating. Ciscoes, members of the salmon family, roam in the middle of deep lakes, below the thermocline but well above the bottom, feeding on algae and zooplankton as well as smaller fish. If you have ciscoes in your lake, be grateful for their contribution to healthy game-fish populations.

Rock bass: The bass that isn't

Rock bass are present in many northern lakes. They are aggressive feeders. They are often found in the same habitat as smallmouth bass. One thing they are not is bass. They are actually members of the sunfish family, with bluegills and pumpkinseeds.

Rock bass, sometimes called rock sunfish, goggle-eye, and redeye, look something like a cross between a bass and a bluegill. They're thicker through the body than bluegills, though not as thick as a bass. The general body shape is longer than a bluegill's, but more oval-shaped than a bass. The mouth isn't small and roundish like a bluegill's; it resembles that of a bass, though it's not proportionally as big. Like their panfish cousins, rock bass have a two-part dorsal fin, a spiny portion fore, and a soft-rayed portion aft. Perhaps the most striking thing about rock bass is their coloring. The eyes are usually bright red. The belly is white; the body ranges from golden brown to olive with rows of black dots. The sides often have a mottled pattern that can vary greatly from one fish to another.

You'll find rock bass living in groups, mostly in cooler, clearer lakes, in areas with gravel or rocky bottoms. Submerged logs and brush are added attractions. They feed on underwater insects, crayfish, and small fish, including their own fry. Sometimes they rise to pluck bugs off the surface. An interesting quality of rock bass is that they can quickly change color to silvery or blackish to match their surroundings. They can be quite beautiful when lifted from the water.

Suckers: Here's to Hoover-mouth

Most of us know suckers mainly as bait. Raised in ponds, they are sold in different sizes as enticements for walleyes, northern pike, and muskies. If you grew up on a Great Lakes tributary stream, you may have fished for suckers using a dip net hung by a rope from a bridge in springtime when the fish migrate upriver to spawn. There's more to suckers than that, though; they are important to the health of the fishery in many lakes. If you don't you see them in your lake, that's because they tend not to take what anglers offer.

Suckers (technically white suckers) are bottom-feeders and have mouths well adapted to that purpose. The leathery lips aim downward instead of straight ahead, so the fish can cruise along, dining in comfort, casually vacuuming up food like insect larvae, worms, small mollusks and crustaceans, plant matter, and fish eggs from the sediment. In turn, suckers are a vital food for favored game fish. Suckers live in almost any lake and stream in northern latitudes. They do fine in clear, clean waters but also tolerate relatively low dissolved oxygen and so can thrive in turbid urban waterways. Suckers have fine scales. Their sides are dark greenish with a metallic luster. The belly is white, and hence the common name. Adults can grow up to 20 inches long and weigh 2 pounds or more; musky anglers are known to use those at the top of the size range for bait in the fall. So even if you never see suckers on your lake except in your bait bucket, be sure to assign them a little respect. Here's to Hoover-mouth!

The ways of bluegills

Almost any angler enjoys catching bluegills. When hooked, they put up a nice battle, ounce for ounce, especially on lightweight gear. Anglers also treasure the flavor of bluegills, pan-fried or deep-fried, bones-in or fileted. The trouble is that bluegills, while probably as abundant as ever, are getting smaller over time in many lakes.

Bluegills are social beings. They spawn not on scattered individual beds but in colonies of nests, row on row, often dozens in the same place on sandy, gravelly bottom. The largest male bluegills occupy nests toward the center of the colony. Smaller "sneaker males" hover at the edges, wait for the larger bluegills to spawn, and then spread their milt over the area.

Where bluegills on average are getting smaller, one likely reason is that anglers are catching and keeping too many. That doesn't mean they're violating the bag limit. It does mean the bluegills perhaps can't sustain being harvested right up to the legal limit by multiple anglers. Fisheries research shows that anglers generally would rather catch fewer, larger bluegills than bigger numbers of small fish. Fewer and larger fish also put more meat in the fry pan. It takes about twenty-five 6-inch bluegills, but only about three 9-inch bluegills, to yield 1 pound of filets. Even if your lake doesn't have a low bag limit, you can choose to keep fewer fish in hopes of helping to improve the size structure. One thing to consider is to let the bluegills alone while they spawn, since that is when they are extremely vulnerable. If you find a colony of bluegill beds in shallow water, it can be great fun just to observe them through polarized sunglasses.

Nibblers: Yellow perch

When fishing, it can be hard to tell from the feel on the line or the bobber's behavior which species is biting. Yellow perch are an exception to the rule. Perch generally don't just take in the bait and swim off, as most fish do. Instead, they tend to nibble at it first. If you're fishing with bait on a jig or a plain hook, you'll feel bursts of jiggling or tapping. If fishing a bobber, it will dance on the surface for brief intervals before finally going down when the fish becomes hooked. That means smaller baits are the ticket. Leave most of a redworm or garden worm dangling from the hook and you may get lots of bites but catch few fish; the perch will just nibble away the loose end of the bait. Perch aren't especially boat- or people-shy. If you fish with kids, you can entertain them by dangling a bait and attracting members of a perch school to the surface.

Yellow perch are a prized species here on our northern lakes and, for that matter, anywhere they're found. The rank right up there with walleyes as table fare, and in fact many people consider them superior. For my money they're best when fried skin on and bones in. If fileted, they're still best with the skin left on, just the scales removed. Perch travel in schools and congregate in weedy areas or over beds of sand grass. They prefer fairly shallow water, seldom going deeper than 30 feet, usually staying within a temperature range of 68 to 72 degrees F. On summer mornings and evenings, perch move toward shore to feed; in spring and fall they often feed throughout the day. They are the epitome of omnivores—they'll eat dragonfly, mayfly and damselfly larvae, worms, small mollusks, crayfish, and small fish, including the young of bluegills and sunfish.

Most perch we see in our lakes grow to no more than 10 to 12 inches. In the right conditions they can reach 16 inches and weigh as much as two pounds, but most perch anglers are looking for dinner, not trophies. A dozen or so in the 10-inch class will do just fine.

Dean Hall Photography

FROZEN

·3·

THE LAKE IN AUTUMN

Winter is on the way, but it's still officially autumn. Your lake is perhaps never as unwelcoming as in these days leading up to the freeze. At the same time, it is a place of profound and soothing quiet and austere beauty. When the ice goes off the lake in spring, there's promise in the air. The water warms rapidly. Home and cottage owners install their piers and launch their boats. Fishing season opens. Memorial Day comes around, spring turns into summer. Every day, boats and kayaks are on the water, white smoke rises from lakeside campfires, people sit on piers in folding chairs sipping coffee or cocktails, kids play on swim rafts and water trampolines.

Soon, much too soon, that's over. Labor Day weekend for many is the last hurrah. Then or not long after, seasonal cabins are shuttered, water pipes drained, piers and boat shelters removed, boats put into storage. For several weeks by now the water temperature has been falling, to the point where it becomes dangerous; the water on a sunny day wears a blue that reminds one more of ice than sky. There's little activity on the lake, though a die-hard fisherman now and then can be spotted.

I put on a warm jacket, go down to Birch Lake, and stand at the water's edge. With the leaves down I can see a few more of the homes and cottages, but the shoreline looks more natural, the lake something closer to its elemental state. The algae bloom that occurred earlier in fall is gone now, and the rippled sand on the lake bottom shows clearly. Beds of rushes, brownish gold like straw, line parts of the shore. Birches stand stark white against the background of conifers; the tamaracks, needles mostly shed, display faintly brown skeletons.

Summer's pier flags and wind socks are missing. No ducks or loons ply the water. I've seen an eagle lately, but not today. A few turned-over rowboats and canoes lie on shore. Swim rafts and boat shelters have been dry-docked. What I notice most is the quiet; I hear nothing except wavelets lapping gently against the trunk of a dead birch that toppled back in September.

In a few weeks the lake will freeze and become, once again, inviting, though in a different way. If the snow waits until the ice is a few inches thick, there may be skating. Eventually ice anglers will place their shanties over the prime walleye haunts. The snowmobile trail will open. Otter slides and assorted animal footprints will appear in the snow. That's worth looking forward to, but the lake's beauty in late autumn is not to be missed, or forgotten.

CLOSING TIME

I was among a few fortunate souls who spent a recent October weekend at their homes or cabins here on Birch Lake. I took a farewell paddle around the lake in a canoe on a prototype autumn Saturday afternoon, clear sky, temperature about 55 degrees F, the softest of breezes, the lake's surface smooth, oaks and birches still holding their colored leaves, the air scented like fine brandy. When traveling alone in my red Kevlar Old Town, I take the bow seat and paddle stern first; sitting farther amidships keeps the canoe level instead of nose-up. At this season, it's appropriate to paddle that way; the trip is more about looking back than forward.

You tend to think, as autumn closes down, on what was, rather than what will be. My annual spring canoe reconnaissances are about watching for life in the shallows, spotting painted turtles released from hibernation, following smallmouth bass across the reef on the lake's east end, spying on walleyes hunkered in tangles of sunken brush. On this mid-October foray, there was little life to observe. The fish had gone deep. Several small ducks in a cluster skittered away and up well before I could get close enough for an identification.

I did encounter several neighbors enjoying the day in various ways: one man disassembling a pier, ratchet wrench periodically rasping; another enjoying a drink while seated atop a short stairway of timbers; a woman at the end of a pier with a small black dog that barked at me sharply; a man and wife preparing a pontoon boat for storage; two fishermen in boats working rocky points, presumably for muskies. From now on there will be few days like this. It's hard at such times not to regret the decline of the seasons and to long, far prematurely, for spring. It's too soon to embrace the idea of November's bleakness and then the long winter. So, while enjoying the day's glory, we scan back over the good times of spring and summer past.

As I pulled the Old Town from the lake and tipped it over on shore, for the last time until next year, the couple from three lots down paddled by in their canoe, just two more lake country folks lucky enough to enjoy this day around or on the water.

SMOKE ON THE WATER

Have you ever looked out over your lake on a very cold autumn morning to find what appears to be white smoke hanging above the surface? It doesn't rise very high and it's not very dense; its tendrils only partly obscure the water. What you're seeing is steam fog, also sometimes called water smoke, frost smoke, or steam mist. Steam fog is an interesting name because you might not perceive it as similar to the steam that rises from a tea kettle, or the fog that can envelop entire landscapes and reduce visibility to almost nothing. In reality, though, a similar principle is behind all three phenomena.

We can expect to see steam fog on our lakes while the water is still relatively warm and overnight and early-morning temperatures drop low. Steam fog forms through the simple process of condensation—water vapor in the air comes out of its gaseous phase and forms tiny droplets of liquid water.

Let's look at the dynamics of steam, fog, and steam fog, one by one. Steam forms when water boils and turns to vapor. The steam that does real work—that cooks our food or drives a steam turbine—is actually invisible. You know this if you've ever opened a bag of popcorn fresh out of a microwave oven. The white wisps that form in the air a few inches above the bag won't burn you, but the invisible steam escaping from right at the edge of the bag definitely can if you're not careful. What happens is that the air inside the bag is very hot and very wet; when it escapes it meets much cooler air. Hotter air can hold more moisture as invisible vapor than cooler air can. So when the steam meets the cooler air, it condenses into droplets that we can see. The water condenses around tiny specks of dust (nuclei) always present in the air.

The various kinds of fog we observe—at low points such as in valleys or covering the entire landscape—are also caused by a difference in temperature, in this case warm, moist air passing over land or water that is cooler. As the warmer air cools, it gives up some of its water vapor to condensation.

And that brings us to steam fog, an enchanting sight on crisp fall mornings. As the weather cools, water gives up its heat more slowly than does land. Heat from the water warms the air just over the surface. Cold air sits above it. The warmer, moister, less dense air rises. As it mixes with the cooler air, water vapor condenses. If you look at steam fog carefully, you'll see not a static cloud but wisps of fog rising. The steam fog dissipates as the day gets warmer and temperatures equalize. Next time you hear an overnight frost warning, make a point to look out on the lake early in the morning, there's a good chance you'll see steam fog.

THE LID GOES ON

If you wonder what happens in your lake after the ice forms, the answer is: not a great deal. Sure, fish still bite, some species more readily than others. But in general, things get quiet, still, and dark down there under that translucent, snow-covered sheet. The three inputs that make your lake so very much alive in high summer—light, heat, and oxygen—are much less abundant.

Only cold-blooded creatures spend winter in the water, though foraging otters and muskrats may come and go through near-shore holes in the ice. In temperatures not much above freezing, fish move around sluggishly; reptiles and amphibians stay mostly still or outright hibernate. Aquatic insects winter in the bottom sediments. Except to the extent that it receives inflows from a stream or groundwater springs, your lake becomes a sealed container. Very little oxygen gets in. The deeper the snow cover, the less light can penetrate, and the less oxygen plants produce from photosynthesis.

And vegetative life itself is limited. The rooted aquatic plants have long since died back. The populations of plankton have plummeted. Whatever oxygen was dissolved in your lake's water at the time ice formed steadily declines through the winter. If you're able to look through clear ice to the bottom, you may see places where occasional bubbles of gas issue from the muck and rise until they meet the ice cover. But biochemical activity and life itself slow to a crawl.

Imagine what it's like down under the ice. There's barely a sound. Maybe the noise of a roaring wind penetrates sometimes, but there's no sound of wave action, no splashing as eagles strike carrion fish on the surface, no swirling noises as loons dive down to hunt for fish, no whine of outboard engines, just unbroken silence. On windless days and nights, before the snowmobile trails open, it's a lot like that on the surface too. It's a time to treasure the quiet, to feel life's pace slow down, to enjoy a sort of suspended animation that lasts until spring. If it feels miraculous to see the earth burst forth with life as the weather finally turns warm, how much more so to ponder the way lake life blooms again when at long last the ice recedes.

THE HARD WORK OF FREEZING

In freshman high school science class my teacher led us through a simple but enlightening experiment to show the difference between water and ice. The teacher had us half-fill two glasses with water. Into one we dropped an ice cube, and into the other an equal volume of ice water. We then recorded the temperature of the water in both glasses for half an hour. In the glass that received ice water, the temperature dropped instantly, but then began rising and kept doing so. In the glass with the ice cube, the temperature dropped more slowly, but then bottomed out and stayed down as the ice melted. The teacher then asked: Which would be better for cooling a drink on a summer day?

The answer was obvious. The experiment illustrated a property of water called the heat of fusion—the amount of heat energy that has to be removed from water to turn it into ice. The definition of a calorie is the heat required to raise the temperature of 1 gram of water by 1 degree Celsius. It takes a great deal more calories to turn water into ice than simply to change its temperature. In fact, 80 calories must be removed from 1 gram of water in order to freeze it, and during the process the water's temperature doesn't change. This works in reverse too. It takes the addition of 80 calories to melt a gram of ice. In other words, it takes eighty times as much energy to melt ice as to warm water by 1 degree, or the same energy to melt ice as to warm water from 0 degrees all the way to 80 degrees Celsius (from 32 to 176 degrees F).

Now, think about your lake. As the days and nights get colder, the water temperature drops rather quickly. Overnight lows in November and December are often below the freezing point of water. In spite of that, your lake takes a long time to freeze, and it's because of that heat of fusion. The water gets down near the freezing point fairly fast, but once it does, it has to give up a large amount of heat energy before becoming ice. Finally, as the days and nights keep getting colder, ice crystals form along the shore, then a skin over the shallows, and finally a solid sheet of ice takes hold. So begins a new season and another way to enjoy our lakes.

HERE'S WHY LAKES DON'T FREEZE SOLID

The ice on your lake likely never gets more than two or three feet thick, no matter how long and cold the winter, no matter how many subzero days. The reason is a property of water that's unusual among chemical compounds. Most liquids become denser and contract as their temperature decreases, and they continue to contract until they solidify. Water, on the other hand, stops contracting and begins to expand as its temperature sinks below 39 degrees F. Then, when the water turns to ice, it expands even more, becoming much less dense. That is why ice floats.

As winter goes on, the coldest water stays on the surface, where it's exposed to the frigid air. Ultimately it freezes, and the ice gets thicker as cold weather persists. Now imagine what would happen if water and ice didn't behave in these ways. Suppose that ice were denser than water. In that event, ice at 32 degrees F would sink to the bottom of the lake. Ice would then accumulate on the bottom all winter long, gradually filling up the lake's bowl. In fact, scientists say that if water behaved like most liquids, lakes would freeze into blocks of ice.

That ice then would not melt readily in spring the way the ice on our lakes' surfaces does. Lake ice is exposed both to the warmer air above and to sun-warmed water beneath it. A lake bed filled with solid ice would rely very heavily on the sun's heat to melt it; the lake might never fully thaw.

As it is, a skin (admittedly a thick skin) of ice insulates our lakes through the winter, keeping most of the water in a liquid state so that life in the water and on the lake bottom can survive until spring. So we should celebrate this near-miraculous property of water that causes the trend in density to reverse itself at the magic number of 39 degrees F.

WHY WON'T MOVING WATER FREEZE?

A few years ago Birch Lake froze later than usual. In part that was because of mild days, but another factor was the wind, keeping the surface choppy. I kept thinking it would take just one very cold, still night for the lake to freeze clear across. That night came, and in the morning, looking out from our hill as the day brightened, I saw a nearly unbroken sheet of ice, dusted in places by light snow. I wondered why it took a windless night to make the lake freeze. Why wouldn't it freeze despite the wind and waves? What difference should wind make? Isn't it all about temperature, not motion?

I searched for a definitive answer. One source said moving water won't freeze because the motion of the molecules can't slow down enough to lower the temperature. That idea fails because it confuses the macroscopic motion of the water with the sub-microscopic molecular motion that correlates with temperature. The mere fact of physical motion does not impart heat. It turns out there are two reasons why lake water in motion remains as liquid. One is that wave action continuously fractures ice crystals as they form, keeping solid ice from taking hold. Under these conditions, water can actually supercool, remaining as liquid even below the freezing point.

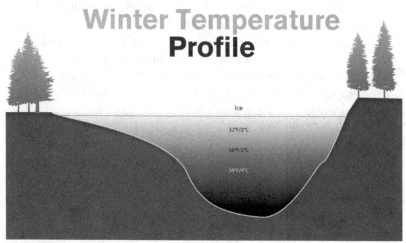

Winter Temperature
Profile

Ice

32°F/0°C

36°F/2°C

39°F/4°C

Eric Roell

The other reason is that wave action on a lake creates a stirring effect, bringing up warmer, denser water from below. Remember that water is densest at about 39 degrees F, and the densest water sinks to the bottom. So the temperature gradient in a lake in winter runs from near 32 degrees F at the surface and gradually up to 39 degrees F right at the bottom. The surface disturbance constantly brings up warmer water, so the surface water can't reach the freezing temperature—until that one still night does the trick.

A critic might then ask what keeps a river flowing all winter, no matter how cold it gets, even if the river is rather slow-flowing and there is no wave action to speak of. The answer is that both factors just described apply—and there is a third: a river is not a closed system. In winter its source is largely groundwater, which in northern latitudes bubbles out of the earth at temperatures around 45 degrees F. That water enters constantly at numerous points along the stream, slowing down winter's cooling and freezing process. Of course, many rivers do freeze on the surface, though the water under the ice keeps flowing.

THE GLASS-BOTTOMED BOAT

A lake is seldom more beautiful than when covered with its first skin of ice, as Birch Lake was one Saturday morning. The first real snow had been followed by a couple of still days and nights. When I woke that

Saturday, I looked out our windows on a surface motionless at last, a light-blue sheen reflecting clear morning sky.

It's a rare treat to go out on your lake's ice while it's fresh and glass-clear. It's rare because it takes those first frigid and windless nights to create a smooth, transparent sheet, and then a few more cold nights to thicken the ice, during which it must not snow. I'm used to seeing the underwater world up close through the lens of a swim mask while snorkeling. Seeing it through new ice is more rewarding because it's an infrequent privilege, and because the water is clearer, the algae having died back, and the silt particles settled. The perspective is different too.

You can view the lake's mysteries at your leisure, as from a glass-bottomed boat. Every feature—weed bed, rock bar, sunken log, trails made in mud by foraging clams—lies exposed. Lying down for a long, close look is an ideal meditation. Skating on clear ice is a whole different adventure, gliding along, watching the silent, sunken world pass beneath you, regretting the scratches you leave behind. If you're lucky, you can spot a fish and follow it at a respectful distance until it takes fright and darts away.

On new ice, it's wise to be careful before venturing out for a look. I like to gauge the ice's thickness by sound before taking the first tentative steps. I find a rock about the size of my fist and throw it as high as I can, on an arc so that it lands a respectable distance from shore. If it breaks right through, I know I have a while to wait. If it makes a hollow *poook* that seems to reverberate along the surface, the ice is still fairly thin but gaining strength. If I hear a sharp *thok!* as if I had dropped a brick on a sidewalk, I know ice is solid.

At that point I'll shuffle out, looking for stress cracks that show the ice's actual thickness. If I stay above water no more than knee- or ankle-deep at first, the worst I'll get if the ice breaks is wet feet. I don't go out over deeper water until I'm certain the ice will hold. I hope one winter, before too long, conditions will conspire again to open a window on Birch Lake's world.

OXYGEN AND TEMPERATURE

Once ice covers the surface, a lake is essentially cut off from the rest of the world. Whatever oxygen is dissolved in the water at that point will only decrease as winter goes on. And yet water creatures routinely survive for four or five months without oxygen replenishment.

That's because, first of all, before the ice set in, the lake's water mixed thoroughly, distributing oxygen more or less evenly throughout. Turbulence stirring the water's surface during October and November ensures that oxygen approaches the point of saturation. Second, because fish and other water creatures in our lakes are cold-blooded, their metabolism slows down to very low levels when the water temperature hovers at or just above freezing. That means they need much less oxygen to live than in summer when the water is warm. The bacteria that during the warmer months break down dead plant and animal matter, in the process using up oxygen, largely go on winter vacation as well. In fact, all the biological and chemical processes that tend to consume oxygen slow down greatly when the water is cold.

But there's another reason lake life can make it through a long winter with, essentially, a finite scuba tank: water can hold more oxygen when it is cold than when it is warm. In fact, it can hold a lot more. Ideally, water at the freezing point (32 degrees F) can hold a little over 14 parts per million of oxygen. For perspective, trout, the most oxygen-sensitive fish in our lakes, need at least 6 ppm. So in ideal winter conditions, our lakes theoretically could begin the iced-in season with a bit more than twice the level of oxygen trout require.

On the other hand, at 50 degrees F, the water can hold about 11 ppm of oxygen. In the height of summer, when the water can approach 80 degrees F, the maximum oxygen saturation is about 8 ppm, a bit more than half the level that is possible at the freezing point. These figures are maximums—we don't see such high levels in our lakes because various natural conditions reduce the water's true oxygen-holding capacity. The point is simply that winter's cold water holds a much bigger oxygen supply than the very same water in other seasons, especially summer.

So fish and other aquatic life go into the winter with the most abundant oxygen supply they will experience all year, at a time when they need that oxygen the least. It's a good scenario for survival and helps explain why winter fish kills in lakes from depletion of oxygen under the ice are much more the exception than the rule.

JUST YOU AND THE OTTER

If you live on a lake, one of winter's pleasures is walking the snow-covered ice until the snow gets too deep, after which you can walk it on snowshoes. You soon find you're not the only one who takes these walks—animals will have left their tracks before you.

Imagine a cover of powder on the ice and a soft snow falling as you embark in your insulated boots. You stay close to shore, because it's a bit too early in the season to trust the midlake ice but also because this is where you'll find most of the hoof and paw prints. Now and then an animal will shortcut across a bay or across the lake proper, but mostly the creatures hug the shoreline, food and cover close by.

Not far on your walk, you come upon a series of sausage-shaped depressions in the snow, each about 6 feet long, paw prints between. These are the slide marks of otters. You know they're fresh because they remain well defined, the edges not even slightly softened by the falling snow. You may not like assigning human qualities to animals, but when it comes to otters, you can't help thinking that here are creatures who know how to have fun. They don't walk or trot along—they run and slide. Yes, they take a few running steps, then flop on their bellies and glide over the snow. A few more steps and glide again.

And so it goes, the tracks continuing as you walk along. The paw prints' orientation shows that you and the otter are heading in the same direction. You keep your eyes forward, hoping to catch a glimpse, since these marks can't be more than a few minutes old. Here and there the trail heads up into the woods, then emerges again on the ice. You never see the otter, just follow its trail halfway around the lake to where it finally enters the woods to stay. On this day, the new snow has cleared the lake's slate; the only tracks in evidence are yours and the otter's. You're glad to have shared the moment.

WHO MADE THOSE TRACKS IN THE SNOW?

All spring, summer, and fall, animals come and go across our woodland and lakefront properties, but we barely notice because they leave little evidence. In winter, though, we easily see their tracks in the snow. Your snow-covered lake is a great place to track wildlife: the trails traverse open space instead of weaving among trees and brush. The only trouble with winter tracking is that it can be hard to identify the actual prints. The animals' footfalls don't leave clear impressions in powdery snow the way they would in mud or soft sand. You need to go by clues such as the track pattern, the sizes of the impressions, and the spacing of the prints.

An essential step in tracking winter wildlife is knowing who is out and about. For example, you won't find bear tracks in snow, since the bears are denned up until spring. Beavers don't hibernate but typically store enough food underwater to get them through the winter, so they're not seen very often.

To identify those trails in the snow, it helps to know four basic track patterns. First are the hoppers, chiefly squirrels and rabbits. Squirrels leave roughly box-shaped sets of tracks, a larger pair (the hind paws) toward the front in the direction of travel. At each hop, the front paws land first, and the rear paws leapfrog past them. Rabbit track sets are similar except that the front paws fall one behind the other instead of side-by-side.

Then there are tracks that proceed in a nearly straight line. Foxes, coyotes, bobcats, and deer share this pattern. Deer hooves commonly exert enough pressure to leave well-defined cloven marks in the snow. As for foxes and coyotes, if there are no clear paw impressions, track spacing can help you tell the difference: 14 to 16 inches for red foxes, 19 to 21 inches for coyotes.

Raccoons, porcupines, opossums, skunks, and muskrats leave pairs of tracks, one behind the other. The sizes can help you differentiate. Otters, fishers, minks, and weasels leave pairs of prints side-by-side, and otters also leave their unmistakable slide marks. When the snow-covered ice is thick enough to be safe, it's fun taking walks on the lake to see who has been out and about.

AMAZING ICE

The solid form of any substance is far different from the liquid, but certain properties of ice are striking. One such property is expansion. Most compounds contract when they cool and expand when they warm up. At the level of chemical structure, that means the molecules pack closer together when the substance is cold and spread out when it is warm.

Water and ice break that pattern. In liquid water (H_2O), each molecule is bonded to three or four others. In ice, each molecule is bound to four others. That means more open space between the molecules, and so an expanded and thus less-dense structure. Ice expands about 9 percent by volume. Stated another way, ice is about nine-tenths as dense as water. This is why only about one-tenth of a floating iceberg shows above the surface.

Once formed, ice, like any substance, expands and contracts with changes in temperature, creating stress and causing fractures. That causes the booming of lake ice, a wondrous sound, sometimes almost musical, sometimes eerie when it echoes across a wide expanse. If you can tolerate a partly open window on a very cold and still night, try lying in bed and listening to the ice boom. You could call it winter's version of listening to the loons.

On the other hand, booms can be a bit menacing. While walking or skating on a lake or river, you may have heard and seen a stress expansion crack sizzle right past and off into the distance. That can give you the urge to dash for shore or lie flat to minimize the pressure on the surface, and so reduce the risk of breaking through. In reality, though, booming and cracking don't mean the ice is weakening.

Ice also has an interesting structure. You don't see it when looking down at your frozen lake, but ice consists of crystals in the shape of hexagons. These crystals grow from the surface downward as the lake is continuously exposed to cold air. The crystals are packed so close together that the ice has the look of one solid sheet. In addition, ice can "evaporate" directly into the gaseous state (water vapor) without first becoming liquid, a process called sublimation. It happens so slowly that it's hard to notice; it can also happen in reverse—water vapor going directly to ice, a process called deposition.

Light and sound behave differently in ice than in air. The speed of light in ice is just over three-fourths as fast as in air. As for sound, if you're standing on the lake ice and hear it boom, sound waves from the booming pass under your feet well before they reach your ears. Sound travels at about 1,100 feet per second through air, but at some 13,100 feet per second through ice. That's about 2.5 miles per second.

The strength of ice can be looked at in various ways. One basis for comparison is fracture toughness, which measures how easily a crack spreads through a material. In this respect, ice is only about one-tenth as tough as window glass. Another measure is thermal shock resistance: how well a material resists cracking with a sudden temperature change, as when dropping ice cubes into a room-temperature drink. Here, ice is only about one-twentieth as tough as glass. Then there's tensile strength—how much force a section of material can withstand when stretched from both ends. The tensile strength of ice is about half that of bricks. Perhaps more relevant to lake ice is flexural strength—how well a material resists bending under a load. The flexural strength of ice is about the same as a white pine board across the grain.

WHY IS IT SLIPPERY?

One of winter's joys is ice skating on a lake or river instead of a rink. Nothing beats the first venture with blades onto brand-new ice, often clear as glass. I've wondered exactly why we can skate—that is, what makes the ice slippery so that steel blades can slide across it. Many of us were taught that we can skate because the pressure of the blades lowers the melting temperature at the ice surface, creating a thin film of water on which we glide. But if that's true, then why can we also slide across ice while wearing flat-soled shoes, which exert much less intense pressure?

It turns out science has rejected the pressure explanation. A 150-pound person standing on ice wearing a pair of skates exerts a pressure of only about 50 pounds per square inch on the surface. That would barely lower the melting temperature.

There are now two other explanations, apparently not mutually exclusive. One is that friction from the skate blade heats and melts the ice and creates the slipperiness. The other is that the ice surface is inherently slippery. This scenario holds that water molecules at the surface of the ice vibrate more because they are exposed to the air and have no water molecules above them to hold them rigidly in place. Therefore, a tiny liquid film remains on the ice surface even at temperatures far below freezing.

Scientists disagree on which theory is correct. For all the advances in scientific knowledge, ice remains a mysterious substance. So, rather than worry too much about the why of slippery ice, we might as well just get out there and glide around!

AN ICY HURDLE FINALLY CLEARED

I could say that after many years I finally succumbed to the allure of ice fishing. Of course, if I did, I would be a liar; allure had little to do with it. I took up ice fishing a couple of years ago on a modest scale, with the sort of excitement I feel when I finally set aside a day to do my taxes. I mean, I live in the north now. I live on a lake. Winters are long. I suppose I might as well try ice fishing. *Sigh.*

Well, I've tried it now and, what do you know, there *is* an allure. You won't find me out there during days with frigid temperatures, but I'll put in a few days with a jig pole when the weather is tolerably warm. If you haven't ice fished on your lake, you might consider trying it. If you know where the fish tend to be during summer, you pretty well know where they'll be in winter. For me, committing to the sport meant getting over a mental hurdle. In summer I can simply grab a rod and box of leeches, get in the boat, and go. In winter it's put on the polypropylene underwear. Bundle up. Grab the ice drill, bucket, slush scooper, bait, jigs, and rods or tipups. Drag it all down to the lake. Slog through the snow to a spot. Drill holes in the ice. Scoop the slush out of the holes. And so on.

But it turns out the process isn't as tedious as I had told myself it was. And after all, it's fishing. The wise have said that it's impossible to fish and worry at the same time. Well, it's also impossible to ice fish, fend off the chills, and worry simultaneously. On water or ice, there's that same complete focus on a single thing that makes fishing like a form of meditation—except on ice the concentration is deeper, because the fish bite so delicately. You have to be alert for just the slightest tap on the rod or twitch in the line.

Getting started doesn't take much of an investment. With a little help from my friends at the sporting goods store, I got my basic gear for well under two hundred dollars. I probably could have bought it used from an online site for much less. I don't have a power auger or a shanty or a sonar flasher, but what I do have is enough for current purposes. I've scared up a few ice fishing friends—it's not an activity I would care to undertake alone.

Still, there's a touch of magic, of wonder, in it. I ponder sometimes what person first conceived of fishing through ice. There's hardly a landscape bleaker than a couple hundred acres of snow over a lake. And yet, you walk on out, bore a hole, drop in a bait, let it sink, and discover that, lo and behold, something down there is alive. It's inspiring. It's thrilling. It's reason enough to become enthused.

HOW MUCH WEIGHT CAN IT HOLD?

How strong is ice? How thick does it have to be on your lake to walk on, skate on, drive a car on? I grew up with a saying, allegedly from someone's grandpa, that "two inches of ice will hold a team of horses." Every authoritative source contradicts that.

Any discussion of when ice is safe must account for the possibility of springs, flowing water underneath (as on a river), snow on top, objects like logs or rocks protruding, recent temperature changes and accompanying freeze/thaw cycles, and the condition of the ice itself. Experts will tell you there is no such thing as "safe" ice—venturing out is always a risk.

Even with clear-blue ice, formed from calm, very cold nights, authorities disagree on how thick is acceptable. Some say to stay off the ice until it is 3 inches thick. Others say less than 2 inches will do in some cases. I have skated on river ice that thin. It was creaky, so to be safe I stayed near the bank where the water was shallow. If you're looking for a little guidance on ice safety, here's how the U.S. Army Corps of Engineers breaks it down for clear, sound ice:

- Less than 1.75 inches: Keep off
- 1.75 inches: One person on skis
- 2 inches: One person on foot or skates
- 3 inches: One snowmobile or a group of people walking single file
- 7 inches: Automobile
- 8 inches: Pickup truck, 5,000 pounds
- 9 inches: Pickup truck, 7,000 pounds
- 10 inches: Larger truck, 14,000 pounds

So that folk wisdom about the team of horses clearly does not apply. And of course you should always use your judgment instead of blindly following some published advice. Here is an old saying to live by: *Thick and blue, tried and true. Soft or crispy, much too risky. If in doubt, don't go out.*

IF YOU FELL THROUGH

Late in the movie *Titanic*, the lead characters slog their way through waist-deep water in the ship's passageways as they try to escape. The flaw in these scenes is that the people show no signs of physical discomfort—only emotional panic—when the reality is they would have been in utter agony from the cold. We see people diving into ice water during various New Year's polar bear plunges, but that's nothing compared to accidentally breaking through the ice, something that happens to some unfortunate souls every year. The polar plungers have the luxury of being able to get out when they want. Not so someone who breaks through in the middle of a lake.

In that event, drowning might not be the most pressing concern. Hypothermia would be. Odds are that if you broke through, you would find yourself holding onto a "shelf" of unbroken ice around the hole you made. And the very cold water would go to work on you immediately. Imagine you fell through the ice on your lake. If water were between 32 and 40 degrees F, you would have fifteen to thirty minutes before becoming exhausted or losing consciousness.

The initial shock of immersion in icy water brings panic and may stress the body enough to cause instant cardiac arrest. People who survived falling through ice have reported that the first contact with the water drove the breath out of their bodies. If their faces are immersed, the first involuntary gasp may take in water, not air. Disorientation can be immediate and complete. A person may thrash helplessly for half a minute or so before being able to think clearly and act.

Besides all that, cold water can quickly numb the arms and legs to the point where they become useless. Severe pain impairs the ability to think rationally. Soon comes hypothermia, followed shortly by unconsciousness and death, unless someone comes to the rescue and administers proper treatment. Normal body temperature is 98.6 degrees F. You start to shiver when your body reaches about 96 degrees F. Amnesia sets in at about 94 F. You become unconscious at about 86 F, and below 80 F, you can't survive at all. So, don't kid yourself that you could easily survive a trip down through your lake's ice. Be extremely careful out there.

CRYSTAL CANDLES

As winter mellows into spring, something almost magical happens to your lake's ice. It's called candling, and it reveals a property of ice that's hidden from us most of the time. It's fascinating, but it also leads to a significant hazard for those venturing out for late-season ice fishing or other adventures.

As the thaw sets in, lake ice changes from what we know as a strong, monolithic structure to a matrix of crystals, arranged, if imagined from above, as mostly hexagons, like the cells in a bee's honeycomb, though by no means as perfect. It's the same basic geometry we see in snowflakes. The ice crystals align vertically, from the top of the ice to the bottom; they are shaped somewhat like candles. Before the thaw, the columns are strongly fused together, and the ice can support considerable weight.

Toward spring, as the ice warms up, it expands and begins to melt. Through the days and nights, the ice goes through many cycles of warming and cooling, creating stress that causes cracks to form. These cracks tend to form at the boundaries between crystals. Additional melting then occurs, further weakening the bonds between the crystals, and the ice's load-bearing capacity drops sharply. In its candled state, ice is often called "rotten." Walking on candled ice, you could easily break through even though the ice remains more than 12 inches thick. That's how much candling weakens the structure. So, for safety's sake, it's wise to stay off candling ice.

Candled ice can be almost musical. If you were to find a thick sheet of such ice driven by wind up onto shore, and if you were to tap at it, candled crystals would tumble off, making a soothing sound a bit like a set of wind chimes. Magical indeed!

TO THE NEW YEAR

Mine are the only footprints in the snow over Birch Lake's ice; no other boot tracks, no otter slides, no hoof marks of deer. A low sun throws my long shadow across a browned-out reed bed and onto the wooded shore, littered with brown needles. The wind at my back sends ghostly strands of powder skimming across the snow surface. As I plod along in my heavy boots, it strikes me: What a wonder, this gift of a glacier.

Even now in the bleakness of winter, the lake is a treasure indeed. This sheet of white, this blank canvas on which creatures will write new chapters each night, to be revealed in daytime, still brings to mind all I enjoyed in the warmer seasons. Passing over what I know to be a rock bar, I realize I could drill a hole there, lower a bait on a hook, and possibly bring in seafood that would fetch $12.99 per pound at the market.

This lake is home to remarkable creatures, some of the most important too small or barely large enough to see without magnification. Through a complex but direct process of succession, the lake's largest dwellers depend on them. It's a self-sustaining system that in turn supports an abundance of wildlife—raptor, waterfowl, mammal, reptile, amphibian. It is also pure joy to be near.

In warmer times we observe, through the trees on our hillside, the shimmer of blue water. In midsummer, the lake is a place to plunge in and feel the day's heat ebb away into the cool that envelops us. To explore with fins, mask, and snorkel is to discover a silent world. Supple plants undulate in the waves or float their leathery pad leaves on the surface. The bottomscape varies from muck to gravel to boulders, to mats of coontail, then back again. At evening, the lake still, the water's surface doubles, at least, the majesty of a sunset, then later the mystery of a moonless night sky.

I ponder all this during my winter walk, while observing what's here to see in the white season. Halfway around my lobe of the lake I come upon tracks that I'm pretty sure are an otter's, and a round hole in the ice where, most likely, the slender creature starts and ends its forays for food. Yes, it's a wonder still, this lake.

This being the day after the winter solstice, the sun departs early, slipping behind trees to the west. As the day darkens and I turn to retrace my steps toward home, I remember how for all its beauty this lake, any lake, is fragile. The things I do, the ways I live, have effects that, however small, may be hard to reverse. Any wonder, any treasure, needs loving care. Especially a lake. It's a thought to carry into the New Year ahead.

LATE ICE

As the winter wanes and the days get warmer, the sun's rays strike more directly, the earth swinging around in its orbit toward the solstice, holding the Northern Hemisphere to the fire. The magnitude of the sun's effect on ice is surprising even on a day with air temperature not too far above the freezing point. Find any dark object on the ice and you'll see a pool of meltwater around it, inches deep, a cuplike depression.

Here a solitary oak leaf, there a lone maple leaf, floats atop or lies at the bottom of one of these pools. The brown leaves, darker than their surroundings, have absorbed the sun's heat and melted the ice around them. Each pool takes the general shape of the leaf—oblong for oak, more circular for maple. It's much the same with leftover rushes that stood above the ice and were blown over by wind. Each slender rush that once lay on the ice now rests in a watery trench two or three times as wide as the rush itself. Such is the power of radiant heat.

Look closely and you may see, here and there, a tiny creature waking to springtime. Just as these days you sometimes spot a fly clinging to a sun-facing door or window, now perhaps a tiny black spider traces the lobes of a leaf in its ice-water pond or scurries over the ice surface. That surface is much different than, say, two weeks ago. Once covered with snow, the ice became exposed with the warmer weather, and now, instead of a hard, monolithic sheet, there's a granular softness. A sweep with the edge of a boot sole scrapes up a small drift of slushy ice particles. In some places the ice still feels sound underfoot; elsewhere your boots sink down so that at times you fear you might break through.

Seen in panorama, the ice no longer bears a smooth, glossy sheen. Instead, wind-driven clouds sweep across a matte finish. Along the shoreline, the ice still mostly connects to land, firm enough to let you walk on out, but in spots a narrow band of open water reveals brown sand below, and in others semicircular openings reveal the presence of springs. The remaining life of this ice depends on the warmth of the days and the hours of sunlight. All we can say for sure is: It won't be long now.

THE LAKE UNVEILED

I had always wanted to see the ice go out from a lake, and finally I did. It wasn't, as I expected, a matter of observing slow changes over a number of days. In fact, it was sudden, much of the process unfolding in little more than an hour. Lake ice melts from the bottom up. First the snow melts off the surface. Then the sun's rays penetrate the ice and warm the water underneath. Warm air above the ice accelerates the thaw, but it's the warming water below that really does the trick.

It was May 6 when I watched the magic happen. Eight days before that, Birch Lake was still frozen stiff. That day and the next two days brought temperatures about 75 degrees F. Then came three more days of below-freezing temperatures, rain, snow, and sleet, before winter's grip finally broke. At that point, the lake ice still looked solid.

May 4 saw a high of about 75 F, as did the next two days. When I visited the lake's shore Monday morning, May 6, ice still covered all except a small area on the far north side. I could see very little change that afternoon when I left for town around two, but when I returned at five, about 30 percent of our lobe of the lake had cleared, the remaining ice forming an irregular pattern, like continents in an ocean.

Then, at about six, a wind brewed up from the east and began, ever so slowly, pushing the ice away. Sitting on our deck, I could mark with my eye a feature on an ice sheet and note its progress relative to the trunk of a tree in our woods. It was a bit like watching the minute hand on a clock, the motion barely perceptible yet unmistakable. Within about an hour, all the ice had blown off to the west, the stirring action of wind-driven wavelets surely speeding up the thaw at the same time. Just like that, our entire end of the lake lay fully open. By morning, the entire lake had cleared and loons plied the water, crying out for what surely must have been joy. After the longest winter or the worst spring I could remember, a new season had arrived. All I could say to that was: Bottoms up!

WHAT A DIFFERENCE A FEW DAYS MAKE

It's amazing how fast things change. One day your lake is an ice desert. A few days later the ice is gone, the race is on: the race to life. The loons don't even wait for full ice-out. As soon as a suitable patch of water opens, they're back, their wails echoing. The lake, freed from its ice insulation, starts warming rapidly, especially when full sun hits the bottom in the shallows.

As the temperature rises, the fish spawning procession begins. It starts with northern pike, seeking out marshy areas as soon as ice melts along the shorelines. Male walleyes begin staging in rocky, gravelly shallows while the water is just a few degrees above freezing; females follow, and activity peaks as the temperature reaches about 45 degrees F. Yellow perch closely follow the walleyes. Their egg strands may drape over plants or sunken tree branches; early-season canoeists can see them in the shallows.

Meanwhile, every creature—amphibian, reptile, mollusk, insect— gets active. Sit on your deck at night and you'll hear the frogs and toads sing, the individual sounds like instruments in an orchestra. This is also a time to watch the ducks migrate through. A pair of binoculars and a field guide can help you expand your vocabulary from "ducks" to buffle-heads, widgeons, mergansers, teal. There's a current of urgency to it all: time is fleeting. These first springtime weeks are a great time of year, maybe the best of times to spend with your lake.

DAYLIGHT SAVING TIME

On the last evening before Daylight Saving Time, the sun has set over Birch Lake. A thin expanse of clouds above the tree line glows red-orange; higher up, rosy pink tints gauzy cirrus spreading almost to the dome of the sky. The temperature reached the high 50s today, and the sun did its work on much of the remaining snow. It has been like this through most of the week. Translucent ice still covers the lake, yet I'm sure the bird I hear in the woods behind me is a robin.

It's too early for this, of course. Who knows what nasty weather lurks between now and May 1, a more realistic estimation of spring's beginning? Still, I'm glad DST is here. For many of us, the worst part about winter isn't the cold or the snow; it's the darkness. As fall turns into winter I miss the long evenings on the lake. Some people don't like DST and would rather just leave the clocks alone all year. I admit I'm not fond of turning time back an hour in October, but I'm glad to nudge it ahead here in March. Winter, after all, is long, and the shortage of daylight makes it drag. After I've had my fill of snowshoeing and hiking and ice fishing, I long for the season's end, not just for the warmth but for the light.

I tend to notice in stepwise fashion the days getting longer. Sitting at dinner in February, perhaps it strikes me that the sky is still red in the west and pale blue above. And then comes March. By midmonth the sun still hangs above the horizon at 6:00 p.m. And so it was that Saturday. The snow had melted from the lakefront steps. I walked down to the lake and stood for a long moment looking out over the ice, scanning the sky, taking in the quiet. As late as this winter was in coming, and as warm as it has been lately, it's not hard to imagine this lake once again as open water, and soon.

This may not be reliably spring, but the leaf-strewn earth beneath my feet is moist and soft and a hint of pine scent drifts in the air. We may still regress into wintry spells, but on average things will get warmer, brighter, better by the day. I take satisfaction in having entered the tunnel of winter and come out once again on the other side. As I gaze across the lake, I look forward to picking up the bedroom clock, spinning the minute hand once around, and moving the world one big, wide step toward spring.

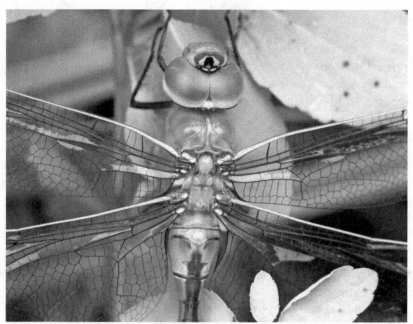

CREATURES

·4·

THE RING OF LIFE

You know your lake is rich in living things, but the land immediately around it is too, especially if it's left natural instead of being planted to lawn. A great deal of life goes on in just a 30-foot-wide swath around the water's edge. Not coincidentally, shoreland zoning laws and regulations tend to protect vegetation in this zone, both to preserve the scenic value of the lakeshore as seen from the water and to preserve the abundance of nearshore life.

The shore zone is inhabited by the cute, slimy, spooky, and beautiful. Cute? How about the water shrew, a tiny mammal, about 3 inches long, that can run on the water's surface and dive under it when not patrolling the land. Water shrews eat mostly insects like stone flies and crane flies, but sometimes they feed on small fish, plant matter, and snails. Slimy? Salamanders spend a major part of their life cycle in the water and emerge to make homes in moist places, like under fallen logs. Cold and clammy is a better descriptor for salamanders than slimy. The same goes for frogs, also abundant in the near-water zone.

Spooky? Bats make their homes along the water, coming out at night to hunt for insects. Beautiful? Loons, of course. They nest right at the water's edge. Dragonflies qualify as beautiful, although some may put them in a "creepy" category. And let's not forget the early morning sighting of a doe and fawn emerging from the woods for a drink.

The lesson is that there's more to our lakes than a pool of water. There's life in it, on it, around it, above it. Next time you head down to your lake, give a moment's thought to the rich zone of life you pass through on the way and make yourself a promise to take special care of it.

ZOOPLANKTON

Those specks you may see in the water as you look down from your pier aren't necessarily algae or pieces of dirt. Populations of zooplankton—tiny cousins to shrimp and crabs that float in the water—are abundant and essential to lake life. They are a vital food source for fish fry, water insects, and the immature forms of frogs, toads, and salamanders. A healthy zooplankton population is critical to a lake's fishery and its overall ecosystem. Scientists often use zooplankton as bioindicators—their populations and even their behaviors can help signal whether a lake is healthy or under stress. Here's a look at the three major forms of zooplankton: daphnia, copepods, and rotifers. Though you may never have seen these three planktonic creatures, and you may never truly see one other than by looking at pictures or through the ground glass of a microscope, it's useful to understand how they contribute to the health of our lakes.

Daphnia: Your lake's fleas

One of the most important living things in your lake is a tiny organism scientists call daphnia and other folks call water fleas. In reality, daphnia

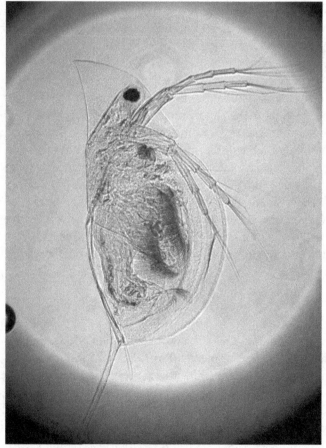

University of Wisconsin Center for Limnology

Daphnia (water flea)

are not fleas, but they're called that because their herky-jerky swimming patterns remind observers of the jumping of fleas, those you hope never infest your dog. You don't need a microscope to see daphnia—the ones in your lake are probably about 1 millimeter in size or somewhat bigger. If you scooped up lake water in a fruit jar and looked through it, you would probably see a daphnia or two kicking about.

They look much more interesting under magnification, with their translucent (actually almost transparent) shell, called a carapace. Through this you can see the innards, including a green gullet if the specimen you're observing has just eaten its fill of algae. The heart lies just behind the head and beats roughly 180 times a minute, about three times as fast as the heart of a healthy human adult at rest. Under a microscope, you can watch the heart beat, watch blood corpuscles pass through the circulatory system, and even see unborn daphnia moving in the brood pouch. Daphnia have helmet-shaped heads that sprout long antennae, which they use, believe it or not, for swimming. A downward thrust of the antennae propels the creature upward; it then floats back down, breathing and collecting food on the way. Steady movement of the ten legs creates a current that moves food into the digestive tract. Besides algae, daphnia eat bacteria and protozoans (one-celled animals). They generally undergo parthenogenetic reproduction—offspring develop from unfertilized eggs. Once hatched, the young molt (shed their shells) several times before becoming adults.

Copepods: Carnivorous teardrops

Copepods have a shape like an elongated teardrop and have a forked tail and a pair of long, curved antennae. They are about 1 to 2 millimeters long and are mainly transparent. There is a single eye in the center of the head. Their name comes from two Greek words: *cope* for oar or paddle, and *pod* for foot. It describes the rhythmic motion of the copepods' appendages as they swim.

And swim they do, the abdomen acting as a rudder that helps them steer. Copepods have two swimming speeds, one using the antennae or mouthparts for slow and steady motion while feeding, and the other characterized by jerky jumping. Here the copepod uses all of its legs at once, in unison, like the oars on a rowing shell. Not all copepods live suspended in the water column. There are several types and numerous species in oceans and freshwater lakes, and some live on the bottom. But those that do suspend are more agile swimmers than the other

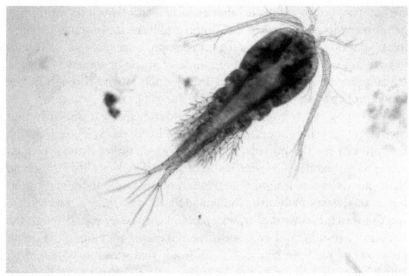

Photo provided by C. Hartleb

Copepod

zooplankton and feed voraciously on algae. For example, one copepod may eat thousands of diatoms (tiny one-celled algae) in a day. They gather food using their antennae and legs, motions of which help create a current of water that carries food toward the mouth. They're not strictly vegetarians; they will feed on almost any living things smaller than themselves, including bacteria and insect larvae such as mosquito "wigglers." They will also consume much-smaller rotifers, attack daphnia, and even take bites from members of their own kind.

In reproduction, males find females by following pheromones—tiny traces of chemicals that leave a scent. The male uses its antennae to hold the female and then places a sperm packet on her abdomen. The sperm enters the female's body through an opening in her reproductive system and fertilizes eggs carried in two sacs. The newly hatched larvae, called nauplius, shed their outside skeleton several times before entering a larval stage in which they resemble the adult copepod. About five molts later, the actual adult emerges.

Rotifers: The wheel deal

Rotifers are arguably the most interesting of these three zooplankton families. Their name, derived from Latin, means "wheel-bearer"—they

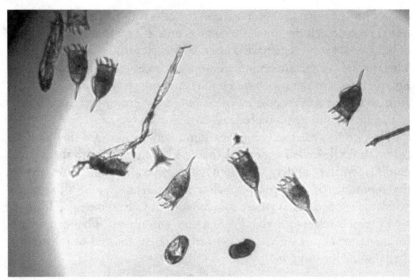

University of Wisconsin Center for Limnology

Rotifers

are sometimes called wheel animals. Rotifers don't really have wheels, but they do have a structure around the mouth called the corona, made up of two tufts containing hairlike growths (cilia) that when in motion give the impression of rapidly spinning wheels. The motion of the cilia creates a current that pulls food into the mouth, where it is ground up by platelike structures (trophi) that act as jaws. In free-floating rotifers, the "wheels" also provide a primitive form of locomotion. Many rotifer species anchor themselves to objects in the water or on the bottom.

These creatures are very small—the most common size range is 0.1 to 0.5 millimeters. That's barely visible to the naked eye; rotifers are best observed with a low-power microscope. They come in a variety of bilaterally symmetrical shapes, some boxlike, others more wormlike. Their bodies are mostly transparent. Rotifers—there are hundreds of species—can be important to a lake's ecology. They support natural water purification by feeding on bacteria, one-celled algae, and organic matter particles suspended in the water. Being present in often large numbers, they are important planktonic members of the food web. They are eaten by larger zooplankton and larval fish.

For such small critters, rotifers have fascinating qualities. For one thing, they contain only a few hundred cells, and they grow in size not by cell division, as most animals do, but by an increase in the size of the cells. That is, rotifers are born with all the cells they will ever have. Rotifers are considered the smallest creatures on earth that fit the description of animals because they have bodies organized into systems of organs. Their organs are greatly simplified forms of those seen in higher animals. For example, a rotifer may have a brain containing fifteen cells, a stomach with about the same number, an excretory system of perhaps a dozen cells, and a rudimentary reproductive system.

Despite this relative sophistication, many rotifers are smaller than common single-celled organisms that live in the water with them. Like daphnia, rotifers undergo parthenogenic reproduction. Some species that reproduce this way consist of all females. Others species also include poorly developed males whose sole function is to fertilize eggs. Rotifers are extremely resilient to changes in their environment. They can survive for many years in a dried or frozen state; when thawed or rehydrated they resume life as usual.

SNAPPING TURTLES

One of my son's favorite fishing moments on Birch Lake had nothing to do with fish. It was the time he spotted a huge snapping turtle not too far from our boat, its head poking out of the water, somewhere between baseball and softball size. We never got close enough to see the shell, but snappers can grow to nearly 20 inches long, shell alone, not counting the menacing head and a tail like a medieval weapon. Snappers aren't seen as readily as painted turtles because they stay in the water, rarely emerging to sun themselves.

We do see snappers during spring breeding season, when they come up from the water to lay eggs, often migrating far inland to find a place with soft sand and sunny exposure. That means now and then we see one crossing a road. In those events, we're advised to stop and move the turtle to the side. But have you ever tried that with a snapper who isn't the least bit interested in moving? Their claws grip the pavement; they have a center of gravity that's almost subterranean.

As kids we handled snappers by grabbing the tail, but if you do that, you've got to hold the beast at arm's length so it can't extend that long neck and clamp its jaws down on your leg. The neck's snakelike attributes and the hiss a snapper gives before striking are reasons for the second half of the scientific name: *Chelydra serpentina*. If I have to move a snapper, I usually find a stout stick and get her to bite it; then I can drag

Paul Skawinski, University of Wisconsin–Stevens Point

Snapping turtle

her to safety. You don't have to worry about a snapper attacking while you're swimming. In the water, they're timid.

At nesting time, female snappers dig a cavity with their hind legs, deposit two to four dozen leathery-shelled eggs inside (sometimes as many as a hundred), then backfill the hole and return to the water. Baby turtles face horrific odds of getting back to the lake their parents came from. Foxes, minks, skunks, and raccoons just love to dig up a nest and have a breakfast of turtle eggs. Young turtles, a bit over 1 inch long, hatch in about 50 to 125 days depending on temperature. They are easy prey for those same predators, as well as crows.

If they reach the water before they die from desiccation (drying out), it doesn't get much easier. Fish and water snakes are just some of their enemies. Those that survive rise to the top of the food chain. They eat just about anything—fish, snails, crayfish, worms, and frogs to name a few, as well as carrion. Snappers you spot on the road may look mean, but they've had a tough time growing up. So if you find one while driving, stop and give her a break. Carefully.

PAINTED TURTLES: HERE'S WHY THEY BASK

One of the pleasures of Birch Lake vacations before we became residents was taking the kids for a paddleboat ride down the shore and around a point to what we called Turtle Bay. There, in the shallows, two logs stuck out onto the water's surface, and on hot days we would find painted turtles basking there, lined up head to tail, big ones, small ones, medium-sized ones. We would paddle up fairly close to the logs and then try to coast in silently, seeing how near we could approach before the painters took to the water—*ploop . . . plop . . . plop . . . ploop . . . plop*, one after another. Once in a while we would see one swimming underwater close to the boat. I would scoop it up with a landing net and let the kids observe it for a while before putting it back.

On these little journeys I often wondered why painters bask in the sun. It turns out that some turtles bask and some (including snappers) don't. That is to say, snappers don't bask on logs or on land but do bask by lying on the water's surface absorbing the sun's rays. Painters, on the other hand, will bask on almost any conveniently accessible nearshore surface, like a log, a rock, or an old pier sagging into the water. When a big white pine toppled just down the shoreline from our place, that became, and remains, a popular basking spot.

Turtles don't bask for pleasure the way humans do (or used to before we became aware of the risk of cancer from ultraviolet rays). They also don't bask out of some conscious decision. Basking is an evolved behavior that confers some survival advantages. Because turtles are cold-blooded, their body temperature fluctuates with the temperature of the water they live in. If it's very warm, so are their bodies, and all the biochemical reactions that go into their metabolism speed up. When the water is very cold, all those reactions slow down—to a near standstill during winter. The function of basking is to raise the body to the optimum temperature for food digestion and metabolism. The turtles spend most of their time in the cool water. When they emerge on a hot day, they stretch out their neck and legs and spread their toes, exposing as much body surface as possible to the sun. They may also absorb heat from the surface they lie on. If there are more turtles than available space on a basking log, smaller turtles may climb up onto the shells of larger ones.

Collecting heat isn't the only benefit of basking. Exposure to sunlight (ultraviolet rays specifically) helps turtles produce vitamin D, an essential nutrient. Basking also dries the surface of the shell and skin, and in turn dries out and kills parasites that may be clinging. The sun's heat can also dry out algae on the shell, keeping it smooth and streamlined for swimming. So though turtles don't sunbathe for fun, they do it, whether they know it or not, for their health.

BATS: NIGHT SHADOWS

If you've stayed out on your pier waiting for dark so you can stargaze, or if you've fished well past sunset, you've seen the ghostly shapes of bats darting over the water. You don't get to see them clearly; they don't come out until it's so dark that all you can see are their fast-moving silhouettes, and then only if there's just enough light left in the sky or reflecting from the water to form a backdrop.

It's no mystery what bats are doing out over the water. They're eating. Many kinds of insects come out over the lake at night: midges hatching from the lake bottom, mosquitoes, moths, and others. Bats vacuum them up by the hundreds, on the wing, at high speeds. Contrary to widespread belief, bats aren't blind, but they don't use their vision to locate prey in the dark. Instead, they use echolocation, an extremely sophisticated form of sonar. Bats send out high-frequency pulses of sound from their mouth or nose and then listen for the echo. From the echo, they can tell the size, shape, texture, and location of the object from which the echo comes back, and then go get it. The process is so precise that bats can detect objects as small as the width of a human hair. Of course, their discernment isn't perfect. A couple of times while night fishing for walleyes, I've had a bat collide with my monofilament line and get stuck there for an instant, as if mounted like a butterfly in an insect collection, before taking off again. I confess that's a bit unnerving.

While imperfect, echolocation is an incredible ability. We humans can make sounds with our mouths and hear the echoes come back, but only if the sound is reflected by some fairly large and distant structure. There's no way we can detect an echo from, say, the side of a building 50 feet away, let alone the echo from a creature like a flying insect. And if we could, how would we determine from which of many insects the sound came? Bats can do this and more.

Just how many insects bats eat is unclear, but the consensus among scientists seems to be that a single brown bat can eat several hundred to even a thousand mosquitoes or similar-sized insects in an hour. That's a lot of pesky bloodsuckers taken out of circulation. Lately, we see far fewer bats than we used to because of white nose syndrome, a fungus that has killed bats by the millions all over North America. Let's hope that scientists can develop a remedy for this condition or that bats develop immunity to it. Without bats, an important bit of magic would be missing from the night air over the lakes we love.

CLAMS AND MUSSELS: NOT JUST STATIONARY OBJECTS

When I was a teenager spending a week with my family at a Northwoods lake cottage, my oldest sister got the idea to harvest some clams and make chowder. Long story short, it was awful. We learned that freshwater clams aren't much good to eat. They are, however, important creatures. Live clams, when closed tight, as they are if you pluck them from the lake bottom, are basically like rocks or skipping stones. But the insides of their discarded shells display a pearly beauty, and their lifestyles are a lot more interesting than their sedentary habits would indicate.

Clams and mussels belong to a family of mollusks called bivalves—their shells are made up of two halves. Mussels' shells are rather flat and oblong; clams' shells are more rounded. Many species of mussels are listed as endangered, threatened, or of special concern because of habitat destruction, poor water quality, and overharvesting. The casual observer might think clams just lie there half-buried in the lake bottom. You know different if you have ever seen the winding trails clams leave behind on their largely random travels. These trails are perhaps best seen in winter through the ice, since there's no water movement then to stir up the bottom and erase them. Clams move around, very slowly, by using a muscular foot that sticks out between the halves of the shell. They plant the foot into the sand or gravel of the bottom and pull themselves along, fraction by fraction of an inch. By moving, they can respond to falling water levels or relocate to more suitable habitat.

Clams and mussels breathe and feed by drawing in and expelling water through two tubes called siphons. The gills have millions of tiny hairlike structures called cilia; the motion of the cilia moves the water in through one siphon and out through the other. Additional cilia filter food such as algae from the incoming water and transport it to the mouth. A difference between clams and mussels is that mussels require the involvement of a fish in their reproductive life cycle, while clams do not. In the mussel life cycle, males release sperm into the water and females draw it in to fertilize the eggs. The larvae grow inside the female, who then ejects them. The larvae attach to the gills or fins of a fish and continue to develop, without harming the fish. They later drop to the lake bottom and grow into adults.

Clams and mussels help keep the water clean by filtering out harmful algae and bacteria, absorbing heavy metals, and sifting silt and fine particles that could harm lake and stream ecosystems. So they're not just more interesting than they look—they're much more important.

DRAGONFLY RIOT

For a couple of weeks dragonflies were everywhere around our place on Birch Lake early one summer. Maybe it was the happy coincidence of a dragonfly hatch with the emergence of late May and early June mosquitoes. All I know for sure is that the air was full of dragonflies, sweeping up mosquitoes like vacuums on wings. Though routinely spotted over land, dragonflies are without question water insects—they come out of your lake after a long metamorphosis. The adult stage we see in the air lasts a couple of months, really just a sliver of the insect's life. Dragonflies mate while on the wing. The female lays her eggs on a water plant or directly in the water. When the eggs hatch in a couple of weeks, nymphs emerge. They don't look anything like dragonflies, but they have one thing in common with the adults: they're voracious feeders.

Dragonfly nymphs eat all sorts of water insects and insect larvae. That includes mosquito wigglers, so dragonflies put a dent in the skeeter population long before they can fly. The nymphs are also quite agile in the water. They swim fast and have a jet-propelled "hyperdrive," ejecting water from the anal opening. In a life stage that can last as long as a few years, the nymphs live in calm water, amid rushes, cattails, and other plants. As they grow, they shed their skin several times. Each in-between phase after the skin is shed is called an instar.

Finally, once fully grown, the nymph climbs up the stem of a plant and emerges from its skin as an adult dragonfly, leaving behind a skin called the exuvia—you may at times have seen one of these clinging to a rush in shallow water. Now the dragonfly is ready for serious eating. Dragonflies are so agile in the air that other insects like gnats, midges, mayflies, and mosquitoes have no hope of escape. The dragonfly uses its legs like a basket to catch bugs on the wing, then feeds its prey into its jaws (mandibles) and crushes it before swallowing. A dragonfly can eat

Paul Skawinski, University of Wisconsin–Stevens Point

Dragonfly newly emerged from nymph form

its own weight in bugs in about half an hour. In mosquito season we can be thankful to see squadrons of brightly colored dragonflies sweeping the air around and above our homes and piers.

FRESH LIVE BAIT

Years ago a friend took me to his favorite little bluegill lake. I brought nightcrawlers; he brought some odd-looking green bugs in a coffee can one-third full of water. Long story short, the big bull 'gills in this lake ignored my worms and went crazy for his bugs. At my pal's invitation, I switched tactics and got in on the bounty. These bugs were the larval forms of dragonflies, commonly though not correctly called hellgrammites. True hellgrammites, not found much here in the north, are larvae of the dobson fly. They're big, caterpillar-like things with a nasty set of pinchers at the head.

If you're old enough, you may remember when bait shops advertised hellgrammites for sale. They were actually selling dragonfly nymphs, but for current purposes we'll call them hellgrammites. It's been years since I've seen any at retail, but I know they live in my lake and probably yours too. If dragonflies buzz around your pier in summer, then your lake has hellgrammites. If you can invest the time and energy to collect them, hellgrammites make perhaps the best bluegill and perch bait in all creation. Hook one through the thorax, hang it from a slip bobber, and if there are fish in the area, they'll bite. It stands to reason fish would take something that, unlike crawlers and redworms, actually lives in the lake.

Though you can't buy hellgrammites, you can get them yourself. A long time ago, I took the frame of an old landing net, stretched some quarter-inch wire mesh over it, and fastened it with pieces of wire. With that tool in hand I waded into some weeds and poked around at the bases of them. Now and then I came up with a hellgrammite or two. A couple of hours' work produced a few dozen. I could see why they're not sold in bait stores—catching them is too labor-intensive. The hunting and catching process, though, is a nice adventure in learning about the creatures that live among the plants in the water.

Rather than seek hellgrammites in your lake, you're better off foraging in ponds with vegetation along the edges and with no fish to whittle down the population. Hellgrammites are most abundant from July through September. Keeping hellgrammites can be a challenge. They're on the cannibalistic side and will eat each other up. They also need to be kept cold. Husbands and wives have been known to have interesting discussions around the keeping of those green critters in the refrigerator. I guess that's an argument for a dorm-sized fridge in the basement.

DAMSELFLIES: GLOW STICKS WITH WINGS

When I was a kid, I saw a colorful creature sitting on the very tip of my fishing rod as I sat out in a boat on a lake where my family used to

Paul Skawinski, University of Wisconsin–Stevens Point

Ebony jewelwing damselfly

vacation. It was about 2 inches long, its toothpick-thin, sunlit body a bright, almost luminous blue, wings nearly invisible. It perched there motionless unless I moved the rod tip, whereupon it lifted off, hovered a moment, and alit again the instant the rod went still.

I thought this was a little dragonfly. I wasn't far off; it was a damselfly, a member of the same order of insects (*Odonata*). Damsel is an apt label—these creatures look and act feminine when compared to their bigger and more boisterous dragonfly cousins. At rest, they hold their four clear membrane wings not out to the sides but folded over the back, tilted upward, front to back. In flight, they don't rattle like dragonflies; they are silent, the wings gently flickering as they glide over the water.

Many damselflies are brilliantly colored, more so than dragonflies, in green, red, blue, or yellow. Like dragonflies, they are water creatures, spending most of their life cycle submerged in a larval (nymph) stage. Their way of feeding gives the lie to their feminine bearing. Dragonflies look the part of the predators they are, in flight their large shapes darting about, at rest their big eyes and strong legs lending a look of menace. Damselflies, though their looks deceive, are also stone-cold killers. As they fly, they use hairs on their hind legs to snare smaller insects, then chew them up. Even in the underwater immature stages, damselflies are predators. The larvae catch other insects by shooting out a long, hinged lower lip called a mask. Damselfly nymphs differ from those of dragonflies in the way they breathe. Most have three gills, shaped somewhat like leaves, at the tip of the abdomen. Dragonfly nymphs' gills are internal, like those of fish.

Adult damselflies have interesting ways of mating. They connect in what's known as the wheel position—you may have seen them flying around this way in tandem. While the female deposits eggs from her abdomen into the tissue of an aquatic plant, the male usually stays attached to her, grabbing the front part of her thorax with claspers at the tip of his abdomen. In a nearly vertical position, he can frighten other males away. Watch for these damsels around beds of rushes and anywhere green plants emerge from the water. If you're lucky, one or two of them may pay a visit while you're out in the boat fishing. They provide a nice diversion while you patiently wait for a bite.

FISHING SPIDERS

If you've ever seen a fishing spider on your pier or in your boat, you might have found it a bit unsettling. I'll admit the fishing spiders I've seen have unnerved me, even though I know they're harmless. For one thing, they're big, both in body and in long, athletic-looking legs. The ones I've seen have measured maybe 1.5 inches long, but they can get quite a bit bigger. Females are about twice as big as males and may cannibalize them as part of the mating sequence.

These spiders are fast. At rest they have the look of a racecar jacked up in back. When startled, a fishing spider can dart the length of your boat in a flash. If you manage to shoo it out or flick it out with a whisk broom, a fisher will run like the devil across the water to a new hiding place. It might even disappear under the water to escape. These spiders don't spin webs to trap prey—they are hunters, on dry land or on, around, and under water. Short hairs that cover the body hold a supply of oxygen (as air bubbles) that allows a fisher to stay underwater for half an hour or more. In addition, each leg has a drop of fluid at its tip that repels water. When they touch the water, the feet float on the surface tension, in a manner similar to a water strider.

Fishers' legs are strong enough to hold relatively large prey, like fish fry or tadpoles. They'll often float on the water, sitting still or swept along by a breeze, waiting for prey. Their eight eyes give them excellent vision: they can detect even small movements, see well even in dim light, and have excellent depth perception. Their sense of touch lets them detect insects that fall to the water and are trapped and struggling in the surface tension. They'll also dive as deep as 6 inches or climb down the stalk of a cattail or other plant to chase food. They hook prey with their front feet and inject venom to kill it before eating. That's not creepy at all, right?

Fishing spiders, also called wharf spiders and dock spiders, are quite common, so your odds of seeing one on your lakefront are quite good. The best way to deal with them is coexistence. They won't attack. You could get something akin to a mosquito bite if you picked one up, but why would you? So, run that fishing spider off if you must, but don't kill it.

FROGS: PLAYERS IN THE BAND

I think of spring frogs more as a band than a choir because, to my ear anyway, their calls sound more like instruments than voices. The frogs certainly don't waste time starting the music once the ice is gone. If you live on a lake, and if it has shallow back-in areas or close-by ponds, you'll hear the calls most evenings. Maybe it will be just one kind of call—many frogs all making the same sound. It's pretty easy then to know who is calling. When you hear a variety of sounds, all mixed together, that's when it gets interesting. It's a bit like going to your child's very first band concert and trying to pick his or her clarinet or flute or trumpet out of the noise. Here it helps to be familiar with the frogs' sounds individually. Without that knowledge you will never hear the subtle threads of the music. In the name of boosting your spring music appreciation, here are descriptions of a few spring callers' sounds.

Northern spring peepers' calls are easy to identify because they essentially say their name. The signature call is a short peep, the pitch ascending slightly. If not attuned, you could mistake this for a bird call. The boreal chorus frog's call is often described as similar to the noise from running a thumb along the teeth of a hard-plastic comb. The northern leopard frog issues a deep, rattling snore that intertwines with chuckling, or a sound like a thumb rubbing against a balloon. It's not exactly musical. Eastern gray tree frogs don't carry much of a tune either. They make short, slightly raspy croaks; a bunch of these in a pond, all calling, make a bizarre noise indeed.

Though the American toad is not a frog, its call is also commonly heard in the spring. It's a long trill that can last up to half a minute. Different toads may emit calls in a variety of pitches.

If you want to identify calls, it's better to listen to them. For that purpose, you can easily look for websites where you can hear the calls of common frogs. Spend some time on one of these sites and you'll become a much more sophisticated listener. A short spell on the deck in the evening listening to the frog music makes a pleasant interlude. And as a bonus you won't feel compelled to ask that question common to all parent attendees at grade school band performances: Why are all those *other* kids playing off-key?

TADPOLES AND POLLYWOGS

Toads and frogs tend to reproduce in ponds, but some do so in our lakes. Perhaps you have seen green tadpoles with bodies almost the size of quarters. These of course are frog predecessors. The much smaller black ones, which grow up to be toads, are almost exclusively seen in ponds. In summer you may have seen tiny toads, barely over half an inch long, hopping on your driveway or on a quiet road.

Frog reproduction varies a little from one species to another. Tree frogs, for example, go through their metamorphosis almost solely in ponds and other standing water. Northern leopard frogs sometimes reproduce in lakes—lake dwellers may see them darting aside as they walk through the grass on their properties. Tadpoles of some frog species grow to be much larger than those of toads before making a transition to life as air-breathers. Leopard frogs, for instance, emerge from the water as adults between 1 and 1.25 inches in length. Bullfrog tadpoles may reach 4 to 6 inches before they become frogs. Species also differ in how long the metamorphosis takes from tadpole to adult. Leopard frog tadpoles make the transition in about three months, give or take, depending on temperature, food abundance, and other factors. Bullfrogs may live as tadpoles for up to two years.

As part of the metamorphosis, tadpoles shift from breathing with gills, extracting oxygen from the water as fish do, to having lungs and breathing on land. Tadpoles first sprout hind legs, and then front legs.

The internal organs then need to change to prepare for life on land. The digestive system has to adjust from a water diet of algae and plant matter to a land menu of insects and worms. Finally the tail vanishes, not because it falls off as some believe but because it shrinks and gets absorbed into the adult frog body.

You may have wondered if there's a difference between tadpoles and pollywogs. I grew up believing that pollywogs were toads and tadpoles could be either frogs or toads. Some people say that pollywogs are tadpoles that have developed their hind legs. Neither is true. Pollywog and tadpole are actually different words for the same thing—a developing form of a toad or frog. Keep an eye peeled for tadpoles in shallow, still areas of your lake. They're reminders of yet another everyday miracle that unfolds under the water.

MEET THE MUSKRAT

Maybe at times on your lake you've seen a clump of green foliage slowly swimming its way across the surface. That foliage was being carried in the jaws of a muskrat. You can easily tell even in winter if you have muskrats on your lake: you'll see a conical mound of brown vegetation sticking up above the ice. They build these lodges in the fall in spots where the bank slopes into the water and it's deep enough so the ice won't form all the way to the bottom.

The muskrats enter the lodge from below; inside there's a room in the center, high and dry, with tunnels connecting to other rooms. In winter, one room may house several muskrats. Their body heat keeps the underwater entrance from freezing shut. Muskrats don't hibernate and they don't store food in their houses. They have to feed actively all winter, foraging under the ice for pondweed and for roots and tubers in the sediment. In other seasons, muskrats feed on plants that grow on the water's edge, such as arrowhead, pickerelweed, rushes, reeds, and—their favorite—cattails. Especially if for some reason plant life is scarce, they'll eat creatures like frogs and crayfish, and sometimes carrion. When they find food, they usually carry it to a protected area like an undercut bank before eating.

Muskrats are well adapted for gnawing plant material. The upper and lower pairs of cutting teeth (incisors) continually sharpen against each other. A bit slow and awkward on land, muskrats are skilled swimmers if not especially speedy, topping out at around 3 miles per hour. They paddle with partly webbed hind feet and use their narrow, laterally flattened and hairless tail as a rudder. When swimming, they can seal off their ears and nose to keep the water out. They can stay underwater for as long as fifteen to twenty minutes to forage for food or get away from an enemy, like a fox, hawk, or owl.

Muskrats tend to stay fairly close to their lodges or to the burrows they dig in banks. At first glance, given their brown color, they can be mistaken for beavers, but of course they are about half the size, about 18 to 24 inches, including 8 to 10 inches of tail. Their web-footed tracks and tail-drag marks in the mud along the shore are distinctive. We call these critters muskrats because they have musk glands under the skin at the base of the tail; these secrete musk on logs and around lodges and banks during the breeding season. And speaking of breeding, they are prolific. A female in a northern latitude usually produces two litters of four to eight young after a gestation of about one month. That could mean a lot of swimming foliage.

WATER STRIDERS: ROW, ROW, ROW YOUR BUG

As kids, my friends and I called them "water spiders." We were wrong; they were actually water *striders*, and they were not spiders but insects. They are nigh impossible to catch—they move on the water's surface with incredible agility. Most interesting is *how* they move. They have six legs like other insects, but they use them differently. The front pair catch prey. The middle (much longer) pair act like oars, for propulsion. The hind pair provide steering, like a rudder.

One might wonder how they stay afloat as they do, legs in contact with the water, body elevated like that of a four-wheeled crop sprayer going over a field. It has to do with the surface tension of water. Surface tension keeps the legs from sinking below the surface. One thing you notice if you watch striders in shallow water are the four roundish shadows the front and hind "feet" cast on the bottom. Those are from the dimples the feet make in the "skin" of surface tension. The surface tension bends, but doesn't break, under the striders' weight.

Closer examination shows that striders' high flotation is not all about surface tension. The legs are covered with microscopic hairs that trap tiny air bubbles, which provide buoyancy. Scientists say studying and replicating these hairs could lead to the development of better water-resistant fabrics. These needle-shaped hairs, called microsetae, are oriented in one direction and have several layers. They are much thinner than a human hair, and the spaces between them trap air to form an air cushion that keeps the legs dry.

It's a treat just to watch these creatures and, if your eyes are quick enough, capture the swift, sure rowing strokes by which they move. Water striders prefer to hang out in shade, such as under tree branches. They eat insects in the water, including mosquito larvae, and land-based insects that happen to fall onto the surface. A sharp mouthpart called a rostrum enables them to suck the juices from the bodies of their prey. Striders need to keep moving to avoid being eaten themselves by an assortment of predators. You can find striders on almost any body of water.

CRAYFISH: FRESHWATER LOBSTERS

Crayfish are the closest thing freshwater lakes have to lobsters. As crustaceans, they are relatives of lobsters as well as shrimp and crabs. Some species are common in northern lakes. One of the most common is the northern clearwater crayfish, scientific name *Orconectes propinquus*, distinguished by its blue-tinted pinchers. These crayfish grow to about 4 to 5 inches, not counting two pairs of long antennae (organs of taste and touch) and the pinchers they extend forward.

The head and thorax of crayfish are covered by a thin but stiff shell called a carapace. They have four pairs of jointed legs. The multi-jointed tail, which has five pairs of small appendages called swimmerets, can be folded under the thorax. Crayfish swim tail-first; the tail provides ample propulsion, and they can scoot out of danger with respectable speed.

Crayfish like to hide out under logs, rocks, or sunken vegetation to avoid predators during the day. They do most of their prowling for food after dark, eating mostly dead fish or other dead material. They reproduce in an interesting way in that the females don't just drop their eggs on the lake bottom. Adults mate in open water in late spring or early summer. The female carries the male's sperm in a receptacle on her underside. She then creates a burrow near the water's edge. In the meantime, eggs develop in the ovary. They are released onto the tail's underside; the sperm fertilizes them and they are held in place by the swimmerets. It takes about three weeks for the eggs to incubate. The young, tiny duplicates of the adult stay attached to the mother for a couple more weeks. Meanwhile, mom comes out of the burrow to feed. Out in open water again, the hatchlings drop off and are on their own.

Crayfish provide a compelling example of what can happen when an exotic species gets established in a lake. Rusty crayfish (*Orconectes rusticus*) were most likely introduced to northern lakes by anglers from more southern states who brought them as bait and may have dumped their bait containers into the water after a day of fishing. Rusties are plenty prolific as breeders and have other traits by which they displace native crayfish like the northern clearwater crayfish. Rusties elbow the natives aside in three basic ways. First, they are extremely aggressive and can force the natives out of their lairs beneath rocks and compete more successfully for available food. Second, the natives are vulnerable to predation by fish when forced out into the open during daytime. Their instinct when threatened is to try to swim away. One can imagine that a crayfish fleeing from a smallmouth bass has about as much chance as an old biplane eluding a modern fighter jet. Rusty crayfish, on the other hand, take a claws-raised defensive posture when confronted by a predator fish. Their outsized pinchers, held up and wide-open, give predators pause, though over time fish do learn to attack and eat them. Finally, the rusties can hybridize with northern clearwater crayfish, creating fertile offspring but ultimately leading to the natives' decline.

Those who live on lakes affected by rusties can fully appreciate the need to protect lakes from invasives by obeying laws that restrict live bait, prohibit the transfer of fish from one lake to another, and require the proper inspection and cleaning of boats. It doesn't take much for an invasive species to get a foothold.

MAYFLIES: UP FROM THE BOTTOM

In summer, you may at times find mayflies clinging to your lakefront entry door or porch screen. They've come up from the bottom of your lake to live their short lives as adults. One thing is certain about mayfly hatches: they don't help the walleye fishing. Walleyes and other fish make a feast of them. If you spot mayflies on your lake's surface near dark, you may see them disappear in circular ripples as bluegills slurp them up.

Mayflies belong to the scientific order *Ephemeroptera*, a name no doubt chosen because these flies are ephemeral. The adults, with their triangular wings that propel them in silent flight in a posture that resembles a man with a jet backpack, live only two nights. They have no

Paul Skawinski, University of Wisconsin–Stevens Point

Mayfly adult form

functioning mouthparts because they don't eat—they have no need to. During their brief time in the air above a lake, male and female mayflies mate in swarms. The female deposits eggs as she flies low over the water or dips her abdomen. Some species even submerge themselves and lay eggs underwater.

From eggs, mayflies develop into adults through several stages of molting. Different species have different molting stages, which also can vary with temperature and water conditions. The insects in immature stages swim to the surface or grab onto rocks or plants. There, they molt into winged subimagoes, which fly quickly from the water to nearby plants, or into trees along the shoreline. There they molt again into adults (imagoes) that fly out over the water to begin the mating cycle anew.

Mayflies are the only group of insects that molt after growing wings. In all other insects, winged forms are found only as adults. A typical mayfly life cycle lasts one year. Immature stages have chewing mouthparts and feed by scavenging small pieces of organic matter, such as plant material or algae that accumulate on rocks or other surfaces. Mayflies require water relatively high in dissolved oxygen, which is why they thrive in fast-flowing trout streams. If your lake has mayflies, that's one sign that the water is in decent condition.

MIDGES: THOSE LITTLE WHITE FLIES

After ice-out on our lakes, we soon see hatches of various flies. Some of these are midges, which come in different sizes. Some are pure white and, flying just above the water, they look like mobile bits of cottonwood fluff. Believe it or not, trout fishing enthusiasts actually tie flies small enough to mimic these midges — and catch fish on them — because midges are important to the food chain. Fish and predatory water insects eat them, and the midge larvae help keep the water environment clean by eating organic debris.

One day in June thousands of midges clung to the screens of our lakeside porch. A tap on the screen sent them flying; in a few moments they were back. The next day only a few remained. I've been down on the lake when midges were thick, a swarm hovering around my head, and if I listened carefully I could hear a faint, collective buzzing. These are non-biting midges, from the insect family *Chironomidae*, often called chironomids. Some people call them "blind mosquitoes"; others call them "fuzzy bills" because of the males' bushy antennae.

Midges have interesting life cycles. The adult flies lay gelatinous masses on eggs on the water surface, each holding as many as three thousand eggs, which sink to the bottom and hatch in about a week. The larvae dig into the mud or, in some species (and there are many), build small tubes to live in. They feed on organic matter suspended in the water and mixed with the bottom mud. In many species, as the larvae grow, they turn pink and eventually dark red, at which point they are known as bloodworms. The color comes from hemoglobin, the same compound that makes our blood red and carries oxygen through our bodies. It allows the larvae to breathe in the mud, which is low in oxygen.

After two to seven weeks, largely depending on water temperature, the larvae become pupae. About three days later, they swim to the surface, and adults emerge within several hours. The adults then mate; they live only three to five days and do not feed. In the heat of summer, midges may complete their life cycle in as little as two or three weeks. Fall larvae do not pupate but instead remain in the larval stage until spring. Watch for midges on your lake throughout the summer. Several generations may hatch before the season turns to autumn.

WATER SNAKE: SLIPPERY, STEALTHY, UNNERVING, HARMLESS

You're relaxing on your pier with a cold drink, or sitting in your boat watching a bobber, when along it comes, three feet long, slithering through the water, straight in your direction. If you've had this experience, you may have found it unnerving, but the creature you observed is actually quite benign. It's the northern water snake, which looks like and is similar in length (at 24 to 42 inches) to the water moccasin. But rest assured that venomous species doesn't exist in the northern latitudes. You wouldn't want to reach down and grab a water snake—it may bite or defecate or regurgitate on you. But you're surely not in mortal danger in its presence.

The water snake is found throughout the northern environment, living in lakes but more often in clean rivers. The snake is thick in the body. Its coloring is variable, consisting of a gray, brown, or tan background with brown, reddish-brown, or black blotches, which tend to fade as the snake ages. The underside is white with bright-red half-moon shapes and dark-gray speckles. Baby water snakes are born throughout the summer. Females give birth to as few as half a dozen and as many as sixty young, sometimes over several days. The newborns range from 6 to 10 inches long. To no surprise, water snakes are excellent swimmers, and not just on the surface: they've been known to stay underwater for an hour or more. They eat crayfish, slow-moving fish, frogs, and other water creatures. They sometimes round up fish or tadpoles with their bodies and then eat them, or just swim through a school of small fish, mouth open. They may also eat dead fish. Their natural enemies include birds of prey and raccoons.

Northern water snakes are active mostly during daylight but sometimes hunt at night. They'll bask in the sun on rocks or logs, or lie in low tree branches above the water. If startled, the snake will just drop into the water and swim away. As with many wild creatures, water snakes are more afraid of us than we are of them. Don't bother them and they surely won't bother you. The best thing to do about them is to accept them as neighbors and to enjoy the sight of their slender bodies undulating through the water.

WHIRLIGIG BEETLES: CAN'T CATCH THIS!

The late basketball coach Al McGuire had his own vocabulary. He called a close game a "white knuckler." The point when a game's outcome became clear was "taps city." An opposing team's big center was an "aircraft carrier." And a small and quick point guard was a "water bug."

If you live on a lake, you know that last term is apt, because nothing is quicker than a water bug, more specifically a water beetle, or more definitively a whirligig beetle. You've certainly seen clusters of these oval, shiny black shapes darting on the surface of still water, making tiny, intersecting, *v*-shaped wakes. As a kid perhaps you liked to catch things, like frogs, crayfish, butterflies, and minnows. It's doubtful you ever caught a whirligig beetle, not even with a scoop net. They are just too fast.

These little beetles are rather flat and streamlined for life on the water. They have two sets of compound eyes that let them see both above and below the surface. Their fairly long forelegs are usually held to the front. The middle and hind legs, shaped like short, flat paddles, provide locomotion, though to see these bugs swim with such smoothness and dexterity, you would swear they were somehow jet-powered. And don't imagine that they spend their entire lives skittering around on the surface. Watch closely and you'll observe that they can dive underwater, which they do to escape from threats or to catch prey.

That's right, these little guys are not as gentle as their carefree behavior would indicate. They have chewing mouthparts, and they feed on smaller insects that fall into the water. They'll also eat dead plant or animal matter. They swim underwater with considerable agility and can stay below the surface for quite a spell. That's because they carry an air bubble attached to the tip of their abdomen. They breathe from this bubble the way humans might from a scuba tank. After they surface, they simply replace the spent "tank" with a new one.

Besides being elusive swimmers, whirligigs have another defense against predators: they emit an odor somewhat like that of apples, and it repels some prospective enemies. Whirligigs lay eggs on submerged plants. The larvae crawl around on the lake bottom, feeding until they mature. When fully grown, they climb out of the water and pupate on plants along the shore. The adults that emerge go back to the lake. They hibernate over winter in mud and plant debris; on awakening in spring they form hunting groups. This accounts for the clusters of them you may see near your pier. There are about fifty species of whirligig beetles in the United States and Canada, and hundreds of species worldwide. It doesn't matter exactly which species you see on your lake. They all look and act pretty much alike. Just enjoy their frenetic choreography, and don't even bother trying to catch one.

Paul Skawinski, University of Wisconsin–Stevens Point

FEATHERS

·5·

ANY DAY, ANY SEASON

The wind blew hard one Sunday in February. I worried that it would take down a tree somewhere and knock out our power, as winds often do in summer. I also wondered what it would be like down on the lake, the wind not filtered by our woods. Feeling the blast as I stepped outside, I remembered another windy, chilly, cloudy, miserable day in July. I was tempted to let that Sunday pass by just sulking in the cabin, waiting for more hospitable conditions. Instead, after a long, late-afternoon walk along the town roads, I took the stairway to our lakefront and stood at the end of the pier.

As was typical of the weekend, gray billowy clouds scudded along on the wind, here and there opening on a patch of blue sky. The northwest wind put a forbidding chop on the water. Intermittent rain speckled my glasses. Then it all unfolded. From over the white pines to the southwest came a bald eagle, then from the north, another. The two soared together for a while. Then one peeled off over the trees and the other sailed along the shoreline slowly, into the wind, so directly over my head that I had to pause and look down at my feet to catch my bearings, making sure not to lose balance and topple into the water. The eagle made a couple of long, looping circles, then flew off to a perch in a pine off to the east.

Next, as if on cue, an osprey appeared above the trees along our shoreline, wings outstretched, motionless, driven downwind like a kite broken free of its tether over where I stood. It curled back into the wind and made a few wide circles over the water before swinging back the way it came. Over the reef on the lake's east end it hung in the air, wings beating steadily, just enough to neutralize the wind. It stayed there for at least a minute, then turned back my way again, passing absolutely straight overhead. A few seconds behind came an eagle again, soaring somewhat higher, closing on the space between itself and the osprey until, if armed with a camera, I could have fit both into the viewfinder frame. They parted ways, the osprey east, the eagle west and upward. Both still patrolled the sky as I turned and headed back up the steps.

Somewhere in there lies a lesson about knowing a lake. Visit it. No matter what or when. No matter how ugly the weather or your mood. You never know what rewards lie in store. For me, the rush of wind and waves and the rain in my face would have been enough. The eagle and osprey gave me a memory. Now I've let this windy February Sunday pass by. I took my walk along the roads; I didn't go down to the lake, though I know the snow wasn't deep and the walking would have been easy. I can't help wondering: What did I miss?

HOODED MERGANSER: WITH A "BROKEN WING"

We're privileged here on Birch Lake to have had a nesting pair of hooded mergansers in each of the past three years. I saw both in spring, the male with its showy crest, black with large white patches, the female with a much more understated ruddy crest. One day I saw the female well out from shore with a large clutch of ducklings. It was hard to count from a distance but it must have been twelve or thirteen, around the maximum for hoodies. It reminded me of when I saw a mother and babies last summer: she treated me to the broken-wing ruse.

Stepping out onto the pier I saw the family, this time half a dozen ducklings, just on the deep side of the rushes to my left. The little ones scrambled into the rushes, but the mom, instead of staying with them, flapped and splashed off toward deep water to my right, dragging one wing. Killdeers are famous for this maneuver, performing it of course on land, to lure predators away from their nests. It's more dramatic when a duck performs it on water. As I stood on the end of the pier, the mother duck paused, maybe 50 yards out. I got into the boat to go fishing, my destination in the mother's direction. As I came slowly her way, she danced even farther, staying at a distance, leading me perhaps another 100 yards before the "broken" wing magically recovered and she took off, skimming low, back into the rushes to rejoin her brood.

If you've seen this performance on your lake, perhaps you've wondered, as I do, where this behavior came from and how it evolved. When did the first duck do this and how did it become a trait embedded in the genetic code, to be passed on to new generations? Remember, evolutionary change is supposed to improve the odds of survival. The female hoodie's low-key plumage surely helps keep her from being seen as she raises her broods, but it would seem the broken-wing act actually puts her more at risk. Yes, she retains the advantage of flight, but the act both exposes her to the predator and coaxes it to pursue her. It's hard to see how that improves her own odds of survival.

Of course, evolution is less about saving the individual than perpetuating the species, and with the broken-wing act the mother draws the predator away from the much more vulnerable brood. But then the question is how the young acquire the instinct to perform the trick that saved them. These are interesting questions. We'll likely never know how the broken wing came to be and what keeps it in the hoodies' behavior pattern, but it's a treat to watch.

DUCKLING SURVIVAL

You've seen mother ducks leading broods of ducklings along the shorelines of lakes. If you watch the same family during a summer, it seems the bigger the ducklings grow, the fewer there are. That's because ducklings have to run a gauntlet through the fifty to sixty days it takes them to reach adulthood. They are vulnerable on land or in the water, and the mother duck can't do much to protect them besides make a fuss. It's not just predators that threaten ducklings. They also face danger from weather that's too hot or cold, lack of food, parasites, and diseases.

Ducks Unlimited states that while nesting success is often considered the key in sustaining strong duck populations, duckling survival is also critical. After all, if the ducklings perish, then it doesn't matter that nests were built and eggs laid and hatched. Important as it is, duckling survival is not well understood. Population surveys indicate that from 30 percent to 90 percent of ducklings fail to reach adulthood, the actual rate depending on multiple factors, among the most significant being the abundance of predators. Many creatures prey on ducklings, especially those newly hatched. Their enemies include muskies, northern pike and largemouth bass, snakes and snapping turtles, foxes, raccoons, minks, and feral cats. Birds feast on ducklings too, especially hawks, owls, gulls, herons, and crows. Broods are especially vulnerable when their mothers have to lead them long distances from upland nests to the lakes or wetland ponds where they grow up.

Bad weather can be a big problem for ducklings as well. We've all been told how water runs off a duck's back, but that's not so true of babies. Their downy feathers insulate them well in dry weather but not when they get wet. In cold, rainy, and windy spells, duckling can die from exposure. Large hail can kill them, and so can especially hot days. It's no surprise that ducklings are the most vulnerable in the first week after the eggs hatch. During this time, many perish from predation and hypothermia. One study in Ontario found that ducklings' odds of survival are nine times greater once they are more than seven days old. Research also indicates that ducklings that are larger when hatched are more likely to survive—they are more mobile, more efficient at feeding, and less susceptible to cold.

Of all variables that affect duckling survival, habitat may be the most important. Ducklings survive better in landscapes rich in seasonal wetlands with plenty of open water, undisturbed shoreline vegetation for cover, and abundant water plants that stand above the surface.

EAGLE EFFICIENCY

If you're an angler, you've likely been amazed at the dexterity with which an eagle plucks a floating fish from the water's surface. Several times in the years I've visited and lived on Birch Lake, an eagle that likes to roost in a cluster of tall pines has swept down to claim a fish injured by being hooked deeply. The fish floats to the surface; the eagle seems to wait until my back is turned. Then I hear it: the *whish, whish, whish* of broad, brown wings. I turn in time to see the bright-yellow talons grab the fish and lift it clear with barely a ripple. It's a surprisingly delicate maneuver for so big and powerful a bird.

The mechanics of it are fascinating. The eagle approaches, talons splayed out far forward, ahead of the beak. If it held this posture all through the catch, the feet would slam into the fish at full flight speed and make a big splash. But that's not what happens. For just an instant, the eagle's body continues forward while the legs decelerate to what looks like a near stop. Seeming independent of the rest of the body, the thick legs drop down. The talons close on the fish, then raise up and for a moment trail out behind, holding the fish aloft as the eagle regains altitude. It all happens in one smooth motion; the bird's basic speed changes little if at all. I liken it to a basketball player attacking the rim full tilt for a layup. The arm has to pull the ball back to retard the ball's speed, as otherwise it would slam hard off the backboard.

The eagle's athleticism in plucking its prey is impressive, and a key to that act is the makeup of the eyes. Bald eagles have eyesight several times stronger than that of humans. They can spot small land animals while flying high. How much easier it must be to spot a dead, floating fish from the top of a white pine. Eagles' vision is superior to ours because the anatomy of their eyes is quite different. Their eyes are about the same size as ours—they fill most of the space inside the skull—but they are shaped differently: the back of the eye is larger and flatter, creating a bigger space for images. Eagles' vision is also much sharper than ours.

Paul Skawinski, University of Wisconsin–Stevens Point

Our retinas include an area at the center called the fovea, which has a high concentration of the rod and cone cells that send visual information to the brain. This is where vision is sharpest. Eagles have not one but two fovea areas in each eye. The second processes images from the sides of the field of view.

Eagle foveae also have much more concentrated and many more rod and cone cells than we do—about five times as many. Eagles also can focus more precisely than we do because they can more readily adjust the shape of the lens and cornea in their eyes. A few other features of eagles' eyes enhance or protect their vision. First, they have a third eyelid—a thin, translucent membrane that periodically sweeps the eye from side to side to remove dust, dirt, and droplets and thus sustain clarity. Second, a fairly large bony ridge above each eye helps provide shade against bright sunlight. Third, in front of their eyes eagles have a patch of bare skin. This ensures that feathers do not grow that might obscure the bird's vision or brush against the eyes during flight.

So all these advantages help eagles when fishing. They can prey on live fish but dead, floating fish are easier. That's because dead fish expose their white underside, while living fish are dark on the dorsal side and harder to see against the colors of the lake bottom. It's also because the refraction (bending) of light in water makes it harder for the eagle to pinpoint the actual depth of a submerged fish. Eagles' eyes can't adjust for light refraction; young birds have to learn through experience to correct for that phenomenon.

FISH HAWK

We see bald eagles routinely here at Birch Lake, but only now and then an osprey. It's a treat to see the graceful flight and angular shape of this one-of-a-kind bird. The osprey is truly a breed unto itself. Ornithologists once classified it with the hawks, but they now place it in its own family. Ospreys, sometimes called sea hawks or fish hawks, are found all over the world. They are superbly adapted to catch fish, which make up nearly all of their diet. They'll capture and eat small mammals, reptiles, and birds only when for whatever reason fish are scarce.

Ospreys' vision is ideal for spotting fish in the water. When hunting, they fly about 30 to 130 feet above the surface. Upon spotting a fish, they hover for a moment and then plunge feet first into the water, often submerging completely, in the process grabbing the prey in their talons. Like owls, ospreys have a reversible outer toe. They can orient their talons three forward and one back, or two forward and two back; the latter configuration is superior for grabbing and holding fish. Rough-surfaced pads on the feet and backward-facing scales on the talons help them hold slippery fish during flight. Ospreys typically carry fish head-first, reducing wind resistance.

They are adept fishers. The Cornell University Lab of Ornithology cites studies showing that ospreys caught fish on at least one in four dives and that some are successful up to 70 percent of the time. They typically hunt for about twelve minutes before catching a meal. Eagles sometimes chase ospreys and force them to drop fish they have caught.

Ospreys are brown on the back and wings, but the underside is mostly white, a striking sight when lit by a low sun. The white head has a black eye stripe that extends down the sides of the face. In flight, unlike eagles and hawks, ospreys fly with a bounce that conveys a certain joy or enthusiasm. Seen from below, their wings bent at the joint and bowed downward, they take the general shape of a capital *M*. In North America, ospreys became endangered in the 1950s from the effects of chemical pollutants such as DDT. They have recovered well and now breed in a wide area, sometimes with a little help from humans. They build nests atop broken trees but also on manmade platforms on structures such as utility towers. In fact, such platforms were important in helping ospreys get reestablished. Keep an eye out for one of these expert flyers and anglers in the sky over your lake.

GREAT BLUE HERON

I saw a great blue heron for the first time when I was a kid. It was early morning and, from the screen porch of the lake cabin where my family was vacationing, I watched this tall, stilt-legged bird pick a fish from a makeshift open-top live box where my dad had stowed the previous evening's catch. Many years later I learned that these big, broad-winged, long-necked birds rear their young not somewhere deep in marsh grass but mainly high up in trees, in huge nests built in colonies called rookeries. As if their skill at capturing fish with their stiletto beaks didn't make them interesting enough.

We've all seen great blues around our lakes, flying along the shore with their easy wing strokes; standing tall on someone's pier or raft; stealing through a bed of rushes in a feeding mode, step by methodical step; giving out an *awk!* and taking to the air with a bluster of wings when a boat drifts too close. Surprisingly enough, these big birds weigh only about 5 or 6 pounds. That's because they are "all legs and neck" but also because, like most other birds, they have hollow bones that help make flight possible.

Great blues above all are expert hunters. They have excellent night vision and can forage in the low light of dawn and dusk. They're omnivorous feeders, willing to take fish, amphibians, reptiles, and even small mammals and insects. With their bills they can grab and hold small prey and spear larger fish clean through. Since the neck is highly flexible and has uniquely shaped vertebrae, they can strike—with impressive speed—at quite a distance from where they stand in the water.

It's the nesting behavior, though, that truly fascinates. It seems ridiculous for birds that might stand a bit more than 4 feet tall to nest in treetops, and yet they do. They tend to establish their rookeries within a few miles of where they feed, most times in swamps or on islands, and around forested lakes and ponds. These locations discourage predators (though not eagles, which attack the young), and so do the nest heights, sometimes 100 feet or more off the ground. The male great blues scrounge for nesting material, chiefly sticks, sometimes taken from other abandoned nests. The female then does most of the nest construction, weaving a platform and a cup for the eggs, then lining it with soft materials like moss, grasses, and pine needles. The building process can take anywhere from a few days to a couple of weeks.

The typical nest is about 20 inches in diameter, although nests can be used from one year to the next and get larger over time, to as much as 3 feet deep and 4 feet across. The female lays two to six pale-blue eggs about 2.5 to 3 inches long, then incubates them for about four weeks. Hatched that high in trees, those chicks had better feel pretty confident about flight before they take that first step out of the nest.

KINGFISHER: WHAT'S THAT RATTLE?

The kingfisher's call is more a noise than a song—a welcome noise, but a noise nevertheless. It isn't musical. When I hear it I tend to think of the sound from an engine that's not willing to start, or some machine gears grinding. I look up when I hear it and search the tree line. The members of this family we see on our lakes are belted kingfishers. They resemble blue jays in being mostly blue and white and having a crest, but they are bigger (typically 13 inches tall versus 11 inches) and behave altogether differently.

Kingfishers have an angular shape when perched and in flight. Males have white breast feathers; females have a rust-colored breast band. The birds patrol lake shorelines and riverbanks and spend significant time perched on tree limbs or wires or in other elevated positions, looking for prey, usually small fish, but sometimes tadpoles, frogs, and water-dwelling insects. When a kingfisher spots a fish, it dives into the water and grabs it in its straight, stout bill. Sometimes a kingfisher hovers over the water, looking down, and strikes from the air. Back at its perch, it pounds the fish on a limb until it stops moving, then flips it up, catches it, and swallows it head first. The bird later spits out the scales and bones as pellets, in the manner of owls.

You're likely to hear a kingfisher before you see it. Don't expect to get a close look without binoculars because these birds have a large "no trespass" zone and will fly off if you approach. They lead solitary lives except during mating and nesting and when tending their young. During that spell, from spring into midsummer, males aggressively defend their territories. If you would like to see a kingfisher nest, good luck. They don't nest in trees. Instead, the male and female dig a tunnel 3 to 6 feet deep, slanted upward from the opening, into a sandy bank. They peck to loosen the earth and then use their feet to push the loose dirt out. The process takes from a few days to as long as three weeks. At the end of the tunnel is an unlined round chamber where the female deposits six to eight white eggs. Both parents help incubate the eggs, the female overnight and the male starting in the morning.

The young hatch in a little over three weeks. The baby birds are hatched without feathers, which grow in about a week; their eyes open in about two weeks. The parents first feed them partly digested fish and then, later, whole fish. The young are ready to leave the nest in about four to five weeks, but the parents may keep feeding them for three or more weeks. After that, they are on their own to forage for fish and make their joyful, mechanical noise.

WOOD DUCKS

Here and there, maybe even on your lake, you've probably seen a wooden box or metal cylinder with a fairly large hole in it, mounted on a post and perched above the water. These are wood duck nests. Woodies (as they're often called) are known for being the most colorful of waterfowl but also for the way their young are launched into the world.

Wood ducks are cavity nesters. Unlike most ducks, they nest in holes in trees close to or overhanging water. In the early twentieth century, woodies were in decline from extensive cutting of mature trees. They rebounded with the regrowth of forests and with help from nest boxes built by humans. Now they're quite abundant and can be seen throughout the eastern two-thirds of the United States. Still, they can be hard to spot, since they spend most of their time in heavy cover.

Male woodies are easy to identify for the bright colors they wear during the breeding season — the scientific name, *Aix sponsa*, is Latin for "water bird in a bridal gown." The crest and top of the head are metallic green, the eyes and bill bright red, the wings blue and black, the breast and rump chestnut red. Once breeding is done, the males transition to more muted colors—wings gray with blue markings, distinctive white patches on the neck and face. As in most duck species, females are less striking, mostly grayish brown, darker on the back and lighter along the sides. They have a gray crest and a white ring around the eyes.

Woodies live mostly in wooded swamps but also around beaver ponds, streams, smaller lakes, and quiet bays of larger lakes. With short, broad wings and a wide tail, they can fly deftly through the woods. Strong claws enable them to grip tree bark and perch on branches. Nest cavities need to be fairly large, ideally a couple of feet deep and about 8 inches in diameter. Woodies typically nest about 30 feet above the ground but sometimes as high as 60 feet. The female lines the nest with her feathers. A nest typically contains half a dozen to fifteen eggs, although broods can be larger than that. The eggs hatch after a few weeks; the young are born with feathers and don't spend much time in the nest. They use their claws to climb to the edge of the nest cavity and then drop to the ground, or directly into the water.

The ducklings need shallow water with ample cover for protection against predators and rich in the insects and small invertebrates on which they feed early in life. As they mature, the ducklings switch to mostly plant foods. Adult woodies feast mostly on nuts, seeds, and water plants but also to some extent on insects and small water creatures. As you canoe or kayak around your lake, keep an eye out for woodies in calm, quite, wooded places. In spring it's a real privilege to observe a duckling's first plunge into the world.

COMMON MERGANSERS: TWO PAIR

As my canoe came clear of a shrubby patch on a small point here on Birch Lake, there came an explosion of wings. Common mergansers—two males and two females—shot out of the water and arrowed away. The contrast of colors surprised and delighted me. I'm used to identifying mergansers by the female's slender shape and rusty crest. The male with his green head (when in mating plumage) can fool the unsophisticated, like me, into thinking he's a mallard.

Those of us who spend time on our lakes in spring get to see a variety of ducks pass through on their migration north. Common mergansers breed mostly in Canada and winter mainly in a swath that includes Ohio, Indiana, Illinois, Iowa, Missouri, Nebraska, and Kansas. I've heard mergansers described as early arrivals in the northward migration, though I've noted other species on our lake sooner after ice-out. Until the sighting of the two pair one recent afternoon, I had never seen more than two mergansers together.

Now and then I've had the chance to watch a female diving for fish, which is mainly what mergansers eat. They vanish faster and surface sooner than do loons—they seem to bring to their fishing a greater sense of urgency. The female's crest looks pleasantly unkempt. As for the male, he's pretty easily distinguished from a mallard. He's similar in overall size but more slender. His green head (not crested) isn't as bright as a mallard's. He also lacks the mallard's chestnut breast and white neck ring. The merganser's bill is long and red; the mallard's is yellow.

You'll also easily distinguish male mergansers by their sound. Mallards give out the "quack" of the stereotypical duck. Mergansers don't say a lot but emit a low, harsh "croak." All that aside, while mallards carry the taint of park ponds and domestication, mergansers portray the essence of the wild. I must say those mergansers that rocketed off Birch Lake, boy-girl, boy-girl, were among the best two pair I've ever been dealt.

SWALLOWS: MASTER AVIATORS

Some wild creatures simply seem to live joyfully. Swallows fit that category. You may have observed them over the water of your lake, performing acrobatic maneuvers as they chase down insects for food. They are built for aviation, but they go about it with more than the characteristic precision of pilots. They dart and dance through the air, doing flight tricks as if for the sheer fun of it.

It's a treat watching them when the lake's surface is smooth on a windless evening; they swoop and dive and skim so low they almost brush bellies with their reflections, and in fact sometimes they actually do so. Swallows (most likely these are barn swallows) are not water birds in the manner of herons or kingfishers; they can just as easily catch their insect meals over dry land. Still, they fly over the lakes quite often, and that's where we can best appreciate their agility. A deeply forked tail distinguishes barn swallows from other swallow species. Barn swallows are about 6 inches long, the head and wings deep blue, the throat and forehead rusty, and the chest and underside a pale orange, somewhat lighter-colored in females.

Swallows are found almost everywhere. Those we see here likely spend winters in Central or South America. Years ago the species nested in caves and in holes in cliffs, but they have adapted to nest mostly in and on human-made structures: under the eaves of houses, under bridges, in culverts and, yes, in barns. A male and female build a cup-like nest together, gathering bits of mud, shaping them into pellets, and carrying them to the nest site. They mix the mud with dried grass and line the nest with feathers.

Females lay four to six eggs, and the male and female share domestic duties that include incubating the eggs and the care and feeding of the young. It takes about two weeks for the eggs to hatch; the chicks can fly in three more weeks, though the parents feed them for about another week. A pair may have two broods in a year and may stay together as mates for several years. Barn swallows will appear over the lakes during hatches of aquatic flies, like midges. To get a drink of water, they'll swoop low and scoop some up with their bill. It's fun watching them perform over the water on a quiet summer evening. You have to wonder, as they go through their spectacular yet effortless maneuvers, how many human aviators and airshow stunt pilots they've inspired.

LOONS: CROWN JEWELS OF THE LAKE

The ultimate Northwoods lake includes wooded shorelines, sandy beaches, crystal-clear water, and natural populations of walleyes and muskies. But even with all that, something essential can be missing: a nesting pair of loons can be a lake's crown jewels. Nothing completes the lake experience like a loon nest somewhere alongshore, and later a downy loon chick paddling beside or riding on the back of its mother.

Loon nesting is often a precarious venture. Being ungainly in the extreme on land, loons need to nest right at the water's edge, or better still, on some high and dry place out in the water away from shore. Once loons' eggs are laid, they are vulnerable to predators like raccoons and foxes. Eggs can also be washed off the nest by waves, including boat wake. The newly hatched chicks have their natural enemies too. Survival is far from guaranteed.

Some lake groups try to foster loon nesting by building artificial platforms. If you build one, they may or may not come. A primary reason loons don't nest on a given lake is that their natural nesting areas have been lost to development. In such cases, a nesting platform can be a quick but less-than-effective fix. Platforms can make the nests more visible to predators like crows, gulls, and eagles, and for that matter curious people who come too close in boats and scare the loons off. The best way to bring nesting pairs back is to restore and protect nesting habitat. That includes making shoreline areas more natural, marking no-wake zones near potential nesting sites, and informing lake residents to stay clear.

None of this means a nesting platform is by definition a bad idea. It's a question of right lake and right location. A nest platform is probably not the best answer if loons already produce chicks on your lake once every three years or so, if loons nest successfully on nearby lakes, or if there are natural nesting areas that could be improved. Good candidates for platforms include lakes where water levels fluctuate greatly, or where loons have nested on mainland shores but have lost their eggs to predators.

If you decide on a nesting platform, it should be placed just after ice-out in water 5 to 6 feet deep, far enough from shore to discourage predators, and not near an eagle nest. In some states, natural resources departments require permits for placement of nest platforms, especially on waters in ecologically sensitive or otherwise designated areas. So, evaluate whether a nest platform is a good fit for your lake. If so, get your lake association involved and give it a try.

Floaters

One September on Birch Lake I saw six loons swimming in a group. For me that was a record—the most I had seen together before on this or any lake was five. First there were two loons, and then, seemingly from nowhere, three, four, and ultimately half a dozen, one or two submerging from time to time. They stayed together just for a few minutes and then, as mysteriously as they had arrived, they dispersed—not even one remained.

Maybe you've seen similar loon kaffeeklatsches on your lake. Early September is the time of year when it's most likely to happen. That's too soon for them to be flocking for migration. It's pleasant to think they're just being social. Most of these loons are floaters—males or females who have no mate and no territory. Floaters are usually seen not in groups but swimming alone on small lakes that have no nesting pair, or on parts of large lakes where no breeding pairs have established territories.

Loons can become floaters in various ways. First of all, loons don't mate during their first two or three years and so lead solitary, wandering lives before settling down. A floater you see may also be a loon whose mate has died from predation, disease, or some other cause. A floater also may be a loon that was driven from its mate and territory in a dispute with an invader. We like to think of loons as gentle creatures, but in reality they can be quite aggressive toward each other. They claim territories when they mate, and they don't welcome other loons in — far from it. Four- or five-year-old loons usually make a home in a territory that had been vacant. Older loons that have not been able to establish territories often try to seize one.

This is where things can get violent. If you've seen two or more loons on your lake acting aggressively toward each other anytime from May through July, then one may be a floater competing for a territory. About one-third of territorial battles between male loons end with the evicted male dying. This in turn affects the makeup of the breeding loon population — there end up being more females than males. The surplus females become floaters. These females will readily mate and claim a territory if they can find an available male. The reality is that there are always more female than male floaters. This means males that become available enjoy what in human dating circles would be called excellent odds. They are just about certain to find a willing mate. Finding a territory — that's another matter entirely. Such is life among the floater segment of the common loon population on our lakes.

Cleared for takeoff

On a bitter cold November day I saw them through the living room window, in a frame of white pine boughs, far out on the lake in a perfect row, four white spots on deep blue. I had my suspicion but reached for the binoculars to confirm, steadying by pressing one barrel against the glass. Yes, they were loons, even at long distance their shapes unmistakable, slowly swimming toward me, white breast feathers lit by a low sun. I was surprised to see them after all the cold, in winter plumage for sure, though so far off that even at 8x magnification I couldn't discern the colors clearly.

I worried for them, since the lake's southwest lobe was largely iced over and a crust on the main lake was starting to push out from shore. I had heard stories of loons getting iced in, though it does seem somehow they know enough to leave before it's too late. Loons need a lot of space in which to take off. Just as a jet plane is marooned at an airport with a too-short runway, loons are stuck if there isn't enough water on which to run and flap up to takeoff speed. The qualities that make loons adept divers and hunters—short wings for streamlining underwater and bodies less buoyant than those of other birds (solid bones instead of hollow)—are handicaps when it's time to get airborne.

If loons live on your lake, you surely know the sound they make as they take flight. It's that sound Fred Flintstone's feet made as he ran his stone-wheeled car up to travel speed: *Pat-a-pat-a-pat-a-pat-a-pat-a* . . . And not just a few *pat-a*'s. Loons have to beat their webbed feet over a long distance to lift clear of the water. Ducks? Startle them and they seem to leap right up, airborne in an instant. Loons, on a calm day, might need to skim 600 to 700 feet along the surface. They need less room if able to take off into a wind, which provides lift, and they do aim themselves upwind if they can, without the benefit of the wind sock human pilots use. Once in the air, they fly fast, some 50 miles per hour, though their flight is energy-intensive. Soaring is out of the question; the wings must beat every second.

So there they were out on the lake in the middle of an Arctic cold front, in all likelihood gone by the next morning or maybe even that same evening.

Winter plumage

Many years ago a friend and I took an autumn fishing trip to a northern lake. As we sat in a rowboat anchored in a favorite spot, up from the water popped a strange bird. It was the size and shape of a loon, but it looked completely different, its feathers in plain shades of gray and brown instead of striking black on white. Although I had never seen it before, I assumed, rightly as it turned out, that this was a loon in winter plumage. It's the thing I most remember from that trip, apart from catching nothing and spending an extremely chilly night in a tent, being it was mid-October and a cold front had come through.

Seeing the loon newly attired, I wondered how the conversion happened—if the feathers fell out one by one and got replaced or if the feathers changed in some way. It turns out that once feathers have fully grown, they are essentially dead tissue; there is no way for them to change other than by fading or from wear. So loons do lose their feathers, a process called molting, and replace them with new ones. The feathers

Linda Grenzer

Common loon, winter plumage

Linda Grenzer

Common loon, summer plumage

don't fall out all at once, which would make the loon look like a plucked chicken, but the process is fairly fast.

Loons molt twice a year, first here in breeding territory, during fall, into their winter drabs, and then once more, into their characteristic black-and-white plumage, before they leave their wintering grounds in the south. They arrive back on our lakes in all their black-and-white glory. The benefit of drab winter coloration is that it attracts less attention from potential predators. Loons also vocalize less while in winter feathers.

The molt is a hazardous time for loons. Being heavy in the body and having relatively small wings for a bird their size, they badly need all their flight feathers, so it's fortunate that they lose and replace those feathers quickly. As it is, they are unable to fly for two to three weeks while the new feathers develop. During this time, they also have to expend a great deal of energy growing the feathers. That stress makes them more vulnerable than usual to diseases. It's on the new, drab feathers that loons head for their winter homes, generally in October or November. By the time they return in spring, they are all dressed up for the mating season and in full, enchanting voice.

Music

Many summer sounds go missing come winter. Most of all we lake dwellers miss the calls of loons coming in through the screens on quiet evenings. The reality is that we miss not one sound but four, each with its own variations, because no two loons' calls are exactly alike. It's interesting when hearing two loons wail across a lake to notice the slight difference in the pitch of their notes.

The wail is the call we enjoy most and the one we usually hear on tourism videos. It's typically a three-note call—the middle one the signature drawn-out high note, given by males and females. The wail is a contact call by which loons locate each other and encourage other loons to come closer. It helps adults call in their mates and lets young ones summon their parents. Loons also wail to alert each other to the presence of bald eagles, which are mortal natural enemies.

On the less friendly side, there's the yodel, a series of sharp, high notes. Only males give this call, and it's a territorial declaration. The yodeler warns intruding loons that he owns the territory and will defend it. Each male has its own signature yodel, and if a male moves to a new territory his yodel will change. When yodeling, the loon takes an aggressive posture, crouching down, extending his head and neck and holding his bill just above the water line.

A gentler, subtler call is the tremolo, which sounds a bit like loon laughter. Loons of both genders use it, including the young starting at two or three months old. It's a distress call—a cry for help—and often is made by a loon being driven away by another defending its territory. I've heard loons tremolo when disturbed by people, such as when a boat ventures too close, or during noisy lakeside fireworks displays.

Finally, there's the hoot, a single, lower-pitched note used at close quarters. We don't hear it at a distance. It can be a greeting or just a way of "chatting." An adult loon might hoot to its mate or to a chick. Loon chicks have another kind of call of their own: a peep to urge the parents to find them food (sometimes known as the begging call). We're not likely to hear this call unless we're lucky enough to observe a family at close range (which ethically speaking we generally shouldn't do). Naturalist Aldo Leopold couldn't imagine autumn without "goose music." Here on the lakes, what would summer be without the music of loons?

Paul Skawinski, University of Wisconsin–Stevens Point

GREENERY

·6·

WHY DO WE CALL THEM WEEDS?

On land, weeds are, by definition, plants growing where humans don't want them. They're the dandelions that crop up in urban lawns, the morning glory that infests farmers' cornfields, the burdock that wants to take over gardens. But what of the plants that grow in our lakes? Why do we call them weeds? In many ways they are as essential to lake life as trees are to forest ecosystems, as grasses and forbs are to prairies. It's true that some water plants are invasive (like Eurasian water milfoil) and deserve to be called weeds. And of course some native and generally beneficial plants can prosper to nuisance proportions.

For the most part, though, water plants serve to promote healthy and diverse fisheries and all manner of aquatic life. If you doubt their value, ask anyone who lives on a lake where a once-abundant plant community has been decimated by rusty crayfish. Few would argue that the change has been for the better. Some nutrient-poor (oligotrophic) lakes have good fisheries despite extremely sparse plant populations, but generally speaking, water plants are a big part of any lake's character. In the most basic sense, they come in four categories.

Emergent plants are those we see along shorelines, like cattails, arrowhead, pickerelweed, and various sedges and rushes. These plants, often in dense stands, help limit the effects of wave action on shorelines.

Plants Emergent

Eric Roell

Plants
Floating-Leaf

Eric Roell

The spongy leaf tissue makes excellent nesting material for muskrats and waterfowl. The plants' interlocking root systems help keep the bottom sediment stable and hold the plants in place against waves and runoff.

Floating-leaf plants—chiefly white-flowered water lily, yellow-flowered spatterdock, and watershield—grow in slightly deeper water, sending flexible stalks up from rootlike structures, called rhizomes, buried in the sediment. The stalks grow rapidly in spring; the plants then die back in winter, leaving only the rhizomes and seeds. The leaves, or pads, have leathery surfaces that help hold moisture. Small pores called stomata admit air to a network of chambers, making the leaves buoyant.

Submersed plants are the ones we most often call weeds; in fact, many include "pondweed" in their common names. These deeper-water plants are the kinds anglers look for when seeking places to cast a lure or suspend a bait on a bobber. The plants are quite diverse in height, leaf structure, root system, and life cycle. They also have many things in common. Being buoyant, they don't need sturdy stalks. Spongy leaf tissue latticed with air spaces helps with flotation.

Free-floating plants are represented mainly by duckweed, with one to three leaves, or fronds, less than one-eighth inch in diameter that simply drift on the surface, taking nutrients from the water through hairlike rootlets. Duckweed is a favorite food for waterfowl and marsh birds and supports insects that provide meals for fish.

Plants
Submersed

Eric Roell

Of course, the word "weed" doesn't carry quite the negative connotation in the water environment that it does on land, but if we didn't think of water plants as "weeds," perhaps we would respect and appreciate them a little more.

HOW WATER PLANTS SPREAD

Plants that grow on land have one main way of reproducing. They produce flowers that get pollinated, then grow some form of fruit that contains seeds. The seeds are dispersed in many ways, some simply scattering on the ground, some taking to the air on bits of fluff, some distributed with the help of birds and animals. There are exceptions to this rule. Some plants, notably grasses, also spread by sending out above-ground shoots (runners) from which new plants spring up. Some plants can grow from cuttings of the stem, if planted in hospitable soil.

Water plants have more varied ways of spreading. Most of them have flowers, a few of them showy, like water lily and spatterdock and emergent plants like northern blue flag (from the iris family) and pickerelweed. The many plants that grow submerged have nondescript flowers that may or may not stand above the surface. These flowers develop into fruiting bodies that hold seeds. But aquatic plants are resourceful; most of them don't rely on seeds alone.

Many aquatic plants spread by way of fleshy underground structures called rhizomes—water lily and hardstem bulrush are examples. All summer long, as the plant's leaves make food through photosynthesis, nutrients are fed down to the rhizomes, which store substantial energy and grow in the bottom sediment. After the leaves die back in winter, the rhizomes remain dormant until spring when the water warms. That's when the rhizomes send up new shoots. Some plants can spread quite extensively in this way. The rhizomes tend to be fibrous and tough. Once established, beds of rhizome-producing plants are extremely durable; new growth appears year after year.

Some submersed plants, large-leaf pondweed (cabbage weed) being one example, can sprout new growth from plant fragments that break loose and settle to the bottom. The fragments take root and grow and over time can form new colonies.

Another mode of reproduction is by turions, sometimes called winter buds. These structures grow at the stem tips of plants, forming in autumn with declining day length or falling water temperature. Turions drop to the lake bottom when part of the stem breaks off or when the plant dies and sinks. They lie dormant through the winter, and new plants sprout from them in spring. Turions tend to be rich in starches and sugars that provide food energy when the new plants get established. In fact, water plants grow much faster from turions than from seeds. Some turions can survive for several days out of water. When you consider all the ways in which water plants can spread, it's no wonder they can be so abundant in our lakes.

AMAZING ALGAE

The word "algae" tends to carry a negative connotation. We may call it scum, goo, or slime. We may think of it as the stuff that gives the water a pea-soup color at certain times of year. While it's true that algae, or phytoplankton, can become a nuisance in lakes that are overly fertile, it's generally beneficial and in fact necessary to a lake ecosystem. It's also an extremely diverse form of life in our lakes.

Green algae

There are many types of freshwater algae, but let's focus for now on the large family known as green algae, not the blue-green algae (in reality a kind of bacteria) that can be toxic and causes all sorts of issues when it blooms. Green algae form the bottom of the food chain. They provide shelter and food for the tiny water animals that in turn are eaten by young fish. They also help a lake absorb nutrients, as well as contaminants like heavy metals. There are several thousand species of green algae, though not all of them live in freshwater. They take an incredible variety of forms, but a key trait they all share is the ability to make food through photosynthesis. Their color comes from chlorophyll, the same pigment that makes land plants green.

Some green algae have only one cell and float freely in the lake; some of those have an eyespot and can actually propel themselves in the water using whiplike structures called flagella. Other green algae grow on underwater plants or rocks. Still others have cells that aggregate in sphere-shaped colonies or in long strands called filaments.

You may have observed filaments as underwater clouds that look a bit like cotton candy, or as a green layer lying along the bottom. If you've ever run a fishing lure through a clump of this kind of algae, what you reel in may have the look and feel of slime, but if you examine closely you can see the individual threadlike strands.

Two common types of filamentous algae are *Ulothrix*, which grows from holdfast cells attached to objects on the lake bottom, and *Spirogyra*, which floats freely. *Spirogyra*'s chlorophyll-containing structures are arranged in a spiral pattern that can be quite beautiful when viewed under magnification.

Green algae can reproduce by cell division (one cell splits in two, creating a duplicate) or sexually by the joining of two reproductive cells.

Volvox: Green galaxies

It's interesting how simple creatures in nature form complex societies. Ants, for example, cooperate to build elaborate nests in and above the ground. Honey bees fashion combs of hexagonal cells to feed and raise their young, and they do intricate dances to tell their hive mates where to locate flowers and nectar. Now, considering your lake, what would you think of individual green algae cells arranged in a cluster that is able to move, synthesize food, and even reproduce as a single organism? That's *Volvox*, a form of colonial algae (sometimes called globe algae). It might not exist in your lake — it's pretty much confined to eutrophic waters — and you won't see *Volvox* colonies in your lake even if they are there; they are tiny, barely visible, if at all, to the unaided eye. To

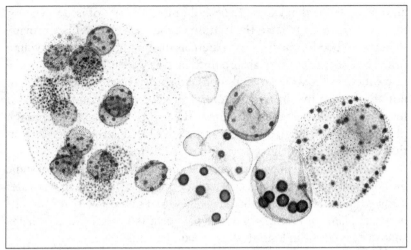

Robert Bell and John Hardy

Volvox colonies

see them you would need to take a water sample and search it with a microscope.

When I see a picture of a magnified *Volvox* colony, for some reason I'm reminded of the Death Star from *Star Wars*. A more appropriate analog, though, would be a galaxy of green stars. A *Volvox* colony consists of about five hundred to several thousand algae cells arranged around the surface of a gelatin-like hollow sphere. Each individual cell has two flagella for mobility, several fluid-regulating organs (vacuoles), a chloroplast that makes food by way of photosynthesis, and an eyespot for detecting light.

The cells are connected by strands of cellular material (cytoplasm) so that they can chemically communicate, enabling the colony to "swim" in a coordinated manner, spinning on its way, propelled through the water by the flagella. The colonies are structured so that cells with large eyespots are clustered at one side to facilitate movement toward sunlight for food production. Specialized reproductive cells are on the opposite side. And reproduce they do—entire colonies, not just individual cells. In fact, *Volvox* can reproduce sexually and asexually. For asexual reproduction, the colonies contain spherical daughter colonies. These emerge as new colonies when the parent colony dissolves. For sexual reproduction there are, believe it or not, male and female colonies, each with different germ cells. The male colonies have sperm cells, and the females have germ cells that enlarge to form an ovum (egg).

The flagella on the individual cells help the colony locate essential nutrients like phosphorus and nitrogen that can be absorbed to promote growth and reproduction. While creating their own food, *Volvox* colonies in turn become food for zooplankton like rotifers and water fleas. In this way they start the food chain that ultimately leads to large predator fish like walleyes, bass, and northern pike.

Diatoms: Where lake life begins

Weeds spring up in your lake as spring transitions to summer, but the most significant plant growth that's happening is not obvious to the eye. As the water warms and sunlight continues to penetrate deep, diatoms are proliferating. These are one-celled algae that multiply profusely in colder water that is high in silica and nutrients that build up over the winter. Diatoms build their cell walls from silica, the main mineral in

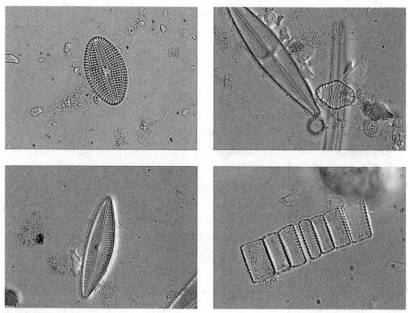

Dr. Krista E. H. Slemmons

Diatoms

sand. As they rapidly proliferate, they often give the water a brownish hue. Unable to control their buoyancy in water, diatoms depend on the action of waves, wind, and currents to stay in depths at which they can receive the sunlight on which their life depends. Without those forces to hold them up, they simply drift to the lake bottom and die.

Later in the season, the diatoms fade and other kinds of algae take over, but a few more words about diatoms are appropriate. For one thing, diatoms, seen under a microscope, are incredibly beautiful—the many species exist in a variety of symmetrical shapes. Diatoms are also an important component of a lake's phytoplankton. Diatoms and other phytoplankton perform the same basic function as grasses in prairies that support grazing animals.

Just like large rooted plants and green alage, diatoms live by photosynthesis; they are among the lake's primary producers. Another function of diatoms is that through photosynthesis they release oxygen. In fact, the diatoms, other phytoplankton, and larger aquatic plants make a net positive contribution to the dissolved oxygen on which fish and other lake creatures depend. So as you watch the water lilies, cabbage weeds, bulrushes, coontail, and other plants pop up in your lake this summer, give a thought to the diatoms, out there by the billions, not doing much besides floating, yet helping to make the whole lake system function.

PONDWEEDS: FORESTS OF THE LAKES

We know the different trees in our woods by their bark and by their shape, but mainly by their leaves or needles. It's also the leaves that help us the most in identifying the many pondweeds in our lakes. In important ways, pondweeds are like underwater forests. Though fully submerged, they stand tall, often brushing the surface. Fish move in and out around their "trunks" and through their "branches." Just as hunters often look for game at the edges of forests, so anglers pursue fish at the edges of pondweed beds.

In the water world, pondweeds serve some of the functions that trees do, providing cooling shade for fish and other creatures, hiding newly hatched and young-of-the-year fish from predators, and producing flowers that in turn yield seeds as food for birds and wildlife. The scientific name for the pondweeds as a group is *Potamogeton*, from the Greek words for river (*potamos*) and neighbor (*geiton*). Pondweeds live in streams as well as in the still waters of lakes and ponds. They can take root in muck or sand, and in shallow water or deep, so long as enough sunlight penetrates to enable growth.

There are many pondweeds, nearly three dozen in the Upper Midwest alone. Some of the more common ones are large-leaf pondweed (called cabbage weed), ribbon-leaf pondweed, floating-leaf pondweed, and flat-stem pondweed. It can be difficult to tell one pondweed from another, but the species share some important features. Unlike rushes that protrude above the water near the shore and floating-leaf plants such as water lily, pondweeds do not have rigid stems. They stand vertical and hold out their leaves to the sun only because their tissues are somewhat buoyant; out of the water they are mostly limp.

Another feature of pondweeds is that the leaves grow on alternating sides of the main stem and have a distinct middle vein. The flowers consist of spikes colored a nondescript greenish-brown; they produce nutlike fruits called achenes. A structure called a stipule grows at or near the base of each leaf. A close look at the leaves and stipules, under slight magnification, can help in pinpointing the species.

Most pondweeds die back during winter and persist by means of rhizomes or turions. Some broad-leafed pondweeds remain alive under the ice. Some can interbreed (hybridize). The same plant may also include different leaf forms, and this can make identification all the more challenging. One thing is clear, though: find pondweed in your lake and you're likely to find fish.

FISHERMEN'S FRIEND: LARGE-LEAF PONDWEED

One of my favorite water plants is large-leaf pondweed. Three of its nicknames—pike weed, bass weed, and musky weed—should tell you why an angler like me would appreciate it. Young fish hang out in it for cover against predators. Perch and bluegills frequent it to feast on insects that cling to the stems and leaves. And so, naturally, the larger predators like largemouth bass, northern pike, and muskies visit it as well. Walleyes use large-leaf beds as a source of shade to protect their sensitive eyes against bright sunlight.

Large-leaf, also commonly called cabbage weed, is rather easy to identify when drifting above it in a boat. It has oval-shaped floating leaves about 2 to 4 inches long. The submerged leaves are among the broadest of any pondweed at about 1.5 to 3 inches; they are about twice as long as they are wide. These leaves are arched and slightly folded; seen from above they look wavy or curly. The plant also has fruiting stalks 1 to 2 inches long, some of which can protrude above the water's surface. The oval-shaped fruits are densely packed on these stalks, which look just a little like the catkins on birch trees. The fruiting stalks appear around midsummer; the fruits are a favored food for waterfowl.

Large-leaf grows mostly in soft lake-bottom sediment in water one to several yards deep, the maximum depth depending on water clarity. Besides reproducing from seeds, the plant can sprout new shoots from rhizomes. New plants can also grow from plant fragments that break loose (or are cut loose by outboard motors) and settle to the bottom. This quality has been helpful for lake groups looking to restore healthy growth of native plants. Some studies show that under favorable conditions, new beds of large-leaf can be created by transplanting cuttings from plants. The flip side to the plant's prolific nature is that in nutrient-rich waters it can grow too abundantly, obstructing boat travel and hindering recreation.

For my part, I would like to see more large-leaf here on Birch Lake. We had large beds of it years ago, and they were great places to fish. When the rusty crayfish took over a number of years ago, they all but wiped out the large-leaf. The beds are coming back slowly since the crayfish have come under control through trapping and fish predation. May the cabbage grow and prosper.

NORTHERN BLUE FLAG: A BEAUTY TO BEHOLD

A flower grows at the base of our pier that would be right at home in a botanical garden, in the hair of a wedding flower girl, or in the corsage of a prom queen. It's best, though, right where it is, blooming in splendor on the shore of Birch Lake, where I see it when I go down to the boat. It looks almost too beautiful to be wild, but it is. It's northern blue flag, a native member of the iris family, and it's quite common around our northern lakes. It can grow in shallow water but more often sprouts from wet sandy or silty soil, in sunlit or partly shaded places at the margins of lakes and ponds.

The plant stands up to 3 feet tall. A fanlike cluster of long, flat, slender green leaves emerge from a shallow rhizome. The flower stalk of each plant can yield three to five blossoms, typically in June and July.

From a distance the flower, about 3 inches in diameter, appears violet blue, but look closely and you'll notice a splash of yellow on white on the gracefully curved petal-like sepals and an intricate pattern of branching purple veins on the white background.

Northern blue flag, a perennial, grows dense roots and rhizomes that help stabilize the soil. Deer don't like to eat it, but muskrats do, and so do various ducks. The showy blooms attract hummingbirds and butterflies, as well as the bees that pollinate the plant. As bees crawl down into the blossom, they rub against the structures that bear pollen, which sticks to their furry bodies. Then they spread the pollen as they travel from flower to flower.

After the blossoms fade, blue flag grows a three-sided seed capsule about 2 inches long. When the capsule breaks, seeds can fall into the water and float away or be carried by wind, helping the plant to spread. The plant also expands its reach through the extension of its rhizomes and so, left to its own devices, it can propagate rapidly in favorable soil.

Northern blue flag is poisonous to humans, especially the rhizomes, which can irritate the skin. However, some Native American tribes used the plant for medicinal purposes, drying the rhizome and using small amounts as a diuretic and a cathartic. Some tribes also used the outermost fibers from the leaves to make strong twine. The roots smell like violets and are sometimes ground to powder, which is added to potpourri and perfumes. All in all, northern blue flag isn't of great practical value to humankind; it's the impractical value that matters. It brightens an early summer morning to walk down to the lakeshore and find the year's first blue-flag blossom, spread wide, greeting the sun, droplets of an overnight rain sparkling on the petals. Most of us would say that's all the value we need.

MILFOILS: THEY'RE NOT ALL INVASIVE

To many lake residents and lake users, "milfoil" is a dirty word. It calls to mind Eurasian water milfoil, a nasty exotic invasive plant that can grow explosively and cause all sorts of problems, crowding out native vegetation and forming mats on the surface that impede swimming, boating, and fishing. In reality, milfoils make up a fairly large group of plants that are, for the most part, native and benign, and beneficial to lake life, as most water plants are. Worldwide there are about seventy species of milfoils, and several exist in the lakes of the glaciated regions of the United States and Canada.

You can identify most water milfoils by the configuration of the leaves. They are arranged around the stems in a whorled pattern—they come off the stem like the spokes from the hub of a wheel, generally four to six leaves in a whorl. (There are some exceptions to this rule.) The leaves themselves are divided like feathers—a central spine with leaflets spreading out to both sides. This gives the plants a graceful, delicate appearance. Milfoils generally grow in water to substantial depths; the actual depth in a given lake depends on how deep sunlight can penetrate. The leaves typically stay below the surface, although flower spikes stand up above the water. Waterfowl eat the foliage and fruits, and stands of milfoil provide cover for fish.

Eurasian water milfoil is believed to have come from Europe by way of the aquarium trade and was introduced by accident to North American waters in the 1940s. The plant doesn't take over in every lake where it invades. Where conditions are right, though, it can spread quickly. For example, stem fragments cut off by boat motors can be carried by waves and currents, settle to the bottom, and take root, after which the plant can spread by way of rhizomes and turions. It also branches out profusely, grows all the way to the surface, and becomes extremely dense.

It can be hard to tell one species of water milfoil from another—it takes a close examination of the stems and leaf and leaflet arrangements. It's important, though, that lake lovers know how to spot the Eurasian variety. If identified early, it's easier to control its spread. If you see a plant in your lake that has feather-like leaves, it deserves close examination. If the plant has long, stringy stems that branch near the surface of the water, and if each leaf has twelve to twenty-one pairs of leaflets, then what you have is Eurasian water milfoil. If that's the case, or if you're not sure exactly what you have, then take a sample and share it with your county's aquatic invasive species specialist for a positive identification.

DUCKWEEDS: FREE FLOATERS

There are water plants in our lakes classified as floating leaf—they have broad leaves that grow on stalks rooted in the lake bottom and lie flat on the surface. Then there's the small group of plants that have no stalks, and no roots as we usually imagine them. Mainly, these are duckweeds, and they simply float wherever wind, waves, and currents take them. Three common species of these plants are small (lesser) duckweed, forked duckweed, and great (large) duckweed. They thrive primarily in warm, quiet bays on eutrophic lakes and ponds. They need water high in nutrients and with a slightly alkaline pH. As the name implies, duckweed is a great food for waterfowl; muskrats, beavers, and some fish also eat it. Mats of it on the surface can inhibit mosquito breeding and provide cool, shady cover for fish. Too much of it can be a nuisance.

You can tell one duckweed from another by the size and shape of the leaf bodies (fronds). Small duckweed has oval fronds about 2 to 6 millimeters long and 1.5 to 4 millimeters wide. Large duckweed fronds are roughly twice those dimensions. Forked duckweed is similar in size to small duckweed but can be distinguished by the fronds' rowboat-and-oars shape. Those differences aside, the species have much in common. The flat fronds contain air pockets that enable them to float. The rootlets of duckweed dangle from the underside of the fronds. Not having the luxury of being surrounded by fertile bottom sediment, they have to draw nutrients directly out of the water. Nutrients can also enter the plant through the frond underside itself.

Duckweeds mostly reproduce by budding. The plant grows a bud that develops into a near-perfect replica; clusters of up to eight plants can form in this way. The new fronds may stay attached to form small clusters.

Though they are small and look delicate, duckweeds are perennials, tough enough to survive through the winter. They form turions that sink to the bottom. With the warming of the water in spring, the turions develop air spaces and float to the surface, where the growth and reproductive cycle begins anew. In high summer, a single duckweed plant can grow one bud per day—it's easy to see how this proliferation can become a problem. However, duckweed can remove large amounts of nutrients and wastes from fish ponds, a highly beneficial attribute.

Duckweed can be grown commercially. It is so rich in protein, essential amino acids, and trace nutrients that it is an excellent food used by fish farmers. It can be directly fed to the fish or dried to form food pellets. Researchers say it even has potential as a source of biofuel, a form of renewable energy. Who knew those green specks floating on the water had so much to offer?

THREE KINDS OF PADS

I used to think all the plants with pads that grew on the water were one and the same species. There were those with big, tough pads, some with yellow blossoms, some with white. Then there were smaller pads, surely just the new and growing stage of the same plant. Of course, I was mistaken. It turns out that three different varieties of water plants with floating leaves inhabit our waters.

The ones with the pure white blooms are water lilies, specifically American white water lilies. They grow long stems that shoot up from a rhizome buried in the bottom. The round leaves (pads) are up to 10 inches in diameter, with a slit that reaches almost all the way to the central stem. The blossoms open early in the morning and close up at about noon. Unlike the leaves of most land plants, water lily leaves have their stomata (pores where carbon dioxide enters the plant) on their glossy upper sides. The spongy leaf stalks have four air channels that carry oxygen to the rhizomes which, by the way, muskrats love eating.

Spatterdock, also called cow lily, has pads similar to those of water lilies, also with a slit, but they are heart-shaped and tend to be bigger. The stems and leaves also grow from rhizomes. The real identifier of

Paul Skawinski, University of Wisconsin–Stevens Point

Bloom of water lily

Paul Skawinski, University of Wisconsin–Stevens Point

Bloom of spatterdock

spatterdock is the bright-yellow flower; when closed up, it looks a bit like a lollipop on a stick. The flowers aren't as showy as the lilies — they don't open up wide the way the white blooms do.

Watershield, also known as dollar pad or water target, is a much more delicate plant. The pads grow no bigger than about 6 inches long by 3 inches wide. Perhaps most interesting, the stems are a little bit stretchy, so in rough water the pads can bob up and down without breaking off. When the plants are young, the stems and pad undersides are coated with a gelatin-like substance that makes them quite slippery. This plant is so named because the pads attach to the stem right in the middle, giving them a shield-like look. The plants sprout little dark-purple flowers in summer that stand just above the surface. Any fisherman will tell you watershield isn't as tough as water lily or spatterdock; if a lure gets stuck in watershield, you can usually pull it free before your line snaps.

HEARTS AND ARROWS: PICKERELWEED AND ARROWHEAD

As the water plants die back in fall here on Birch Lake, one that I miss seeing is pickerelweed. It decorates shallow areas in and around a couple of bays, its glossy leaves and bluish-purple flower spikes standing above the water. Pickerelweed, a perennial, grows in sunlit areas. Its leaves are shaped like the hearts in playing cards (or the spades if you include the leaf stalks [petioles]). The leaves emerge in spring directly from the plant's base and are about 7 inches long and 5 inches wide. When not in flower, pickerelweed can be confused with common arrowhead, whose leaves have a somewhat similar shape and glossy appearance.

Pickerelweed can spread from seed or from rhizomes. The plant can grow in water from just a few inches to as much as 6 feet deep. At maturity, the leaves and flower stalks can stand well above the surface. Pickerelweed is often found growing along with water lily and spatter-dock pads. The flower stalks are pickerelweed's most interesting feature. They range from 2 to 6 inches long. Although the blooms can be seen from early summer into fall, each individual flower lasts for just one day. If you look closely, you'll see that each flower has fine hairs, along with a couple of yellow spots on one petal. The blooms open in sequence from the bottom of the stalk to the top. A single stalk can produce as many as a hundred flowers, densely arranged.

Bees and other insects visit the flowers to feed on nectar and, in the process, pollinate the plant. At the end of each flower's short life, the petals curl inward. Then, assuming the flower has been pollinated, a corky fruit begins to develop. As the season winds down, the flower stalk, now laden with seeds, bends and droops into the water. Wave action can shake the seeds loose and transport them to new locations, helping the plant bed to spread.

Pickerelweed grows densely, helping to stabilize shorelines against erosion. It is excellent for absorbing nutrients from water and improving water clarity, and for this reason many pond hobbyists import and plant it. The seeds of pickerelweed serve as food for ducks and muskrats. They are also, according to many sources, edible for humans. The seeds are nutritious and can be eaten raw, roasted, or boiled like rice. Some people grind them into flour and include them in bread recipes. The young leaves and stalks can also be eaten as greens. (For my part, I'll stick with lettuce and spinach.)

As for common arrowhead (also called duck potato), its leaves are narrower and more pointed than those of pickerelweed. The flowers grow on slender stalks and are much larger and less densely packed than those of pickerelweed. The flowers, white with three rounded petals, are arranged in whorls around the floral stalk. Each flower eventually develops a cluster of seeds—a total of up to twenty thousand per plant.

WHAT'S THE RUSH?

Fishing friends and I have a name for those tall, slender, tube-shaped plants that spike up in clusters in shallow water along lakeshores. We call them pencil reeds. But we're wrong. They are really not reeds at all but bulrushes, most often hardstem or softstem bulrushes. They are emergent plants, a group that also includes sedges, some of which look rather similar to certain rushes.

Both types of plants have solid stems (as distinct from grasses, whose stems are round and hollow), but rushes' stems are cylindrical while most sedges' stems are triangular.

Sedges tend to grow in marshy areas and in the very shallow water of lakes and streams. Several types of rushes grow in deeper water; in the case of bulrushes, out to depths as great as 6 feet. A surprising fact about bulrushes: they technically do not belong to the rush family. They are a kind of sedge. They provide good breeding grounds for aquatic invertebrates and offer cover against predation—and a food-rich pantry—for young fish. The bulrushes have a simple and elegant shape: just beds of green spears poking straight up from the water. They are fibrous and tough, as will be attested by any angler who has tried to drag a lure through bulrushes, only to embed a hook in one of the stems.

Hardstem and softstem bulrushes are similar in various ways. Both grow to be 3 to 9 feet tall, as measured from the lake bottom to the tip of the spear. While the stems of hardstem bulrush are olive green, those of softstem are bluish green. Both plants grow from rhizomes, but hardstem grows better in a firm sand or sand/gravel bottom with good water movement around the roots, while softstem is found in softer, muckier sediment and can prosper in stagnant water.

Despite appearances, these bulrushes are not leafless—leaf sheaths and short leaf blades grow under the water at the bases of the stems, which spring up individually (not in clusters) from the rhizomes. Both bulrush varieties have a drab floral structure at the tip of the stem. Hardstem has flower spikelets that cluster on the ends of rather stiff stalks, while softstem has spikelets on the ends of supple stalks.

Finally, you can distinguish the two types of bulrush by how the stems feel when squeezed between your fingers. The stems of hardstem, to no surprise, are firm, while those of softstem are more spongelike. This is because hardstem has multiple small air chambers in the stem, while softstem has fewer and larger chambers. Anyway, if you see these plants growing in the shallows of your lake, call them pencil reeds if you want, but just know them for what they really are.

COONTAIL CONUNDRUM

With water plants there's often a need to take the good with the bad, to find the happy medium, to see shades of gray. So it is with coontail. It's neither all good nor all bad. It's a native species, but it can spread in proportions that make some want to treat it as an invasive, to remove it with weed harvesting machines or knock it down with herbicides. The coontail here on Birch Lake grows in nothing like nuisance levels, and yet I have seen two sides of it for myself. Snorkeling over beds of coontail, I've observed collections of fish, chiefly perch and panfish, that remind me of scenes from Jacques Cousteau. A favorite place to fish for walleyes is over a coontail patch toward evening. I've also seen through a dive mask how thickly it can carpet the lake bottom, and when I've snagged a baited hook or a boat anchor in coontail, it's amazing what a large clump I can end up pulling from the bottom. It's easy to see how lake residents could be concerned if coontail were to grow out of control. And sometimes it does.

This plant takes its name from the shape of its branches, which resemble the tail of a raccoon, with leaves arranged in whorls around a stem and especially bushy tips. Unlike pondweeds and water milfoils, coontail has no actual roots. It can float freely or anchor itself to the lake bottom by means of stiff, modified leaves. Its dense foliage makes homes for numerous small invertebrates. For that, and for the cover it provides, it attracts fish, large and small. Waterfowl like to eat the plant's leaves and seeds, as do turtles and to some extent muskrats.

But then there can be too much of a good thing. Some people accuse coontail (rightly or not) of providing *too much* cover for small fish, so that they become unavailable to predators, their numbers explode, and their populations become stunted. Excessive coontail growth can limit open water and reduce access for fishing. Coontail isn't particular about how it spreads. It propagates from seeds but also from turions and stem fragments. If uncontrolled, it can crowd and shade out other beneficial plants. When it comes to coontail and some other water plants, an old maxim applies: Everything in moderation, nothing to excess.

WILD CELERY

Did you know that celery might be growing in your lake? It doesn't grow (as best I can tell) here on Birch Lake, but it does in many lakes all over the North Country. Of course this isn't the celery you dice into potato salad, the kind whose stalks you can fill with peanut butter and dot with raisins. It's wild celery, something else entirely. Also called eel grass or tape grass, wild celery grows in areas of hard lake bottom, in water from just a few inches to more than a dozen feet deep.

You can recognize this plant by its long, slender leaves extending up in clusters from rhizomes. The leaves are typically less than half an inch wide, can be up to 6 feet long, and have a broad stripe down the middle. The most notable thing about wild celery is that it's a great food source for muskrats, shorebirds like plovers and sandpipers, and, most of all, waterfowl. Ducks will eat any and all parts of the plant — leaves, rhizomes, seeds, fruit, and especially the tubers (bulbs). Canvasback ducks in particular love this plant. It is often introduced where conservation groups are looking to encourage waterfowl or restore their populations.

Wild celery also helps lakes support fish, as it provides shade and cover for young bluegills and perch, and therefore potential feeding grounds for largemouth bass and other predator fish. This plant has interesting ways of spreading itself around. Separate plants produce male and female flowers. The male flowers, less than one-tenth inch wide, grow in clusters inside casings under the water. Each flower in turn is inside a closed vessel that contains an air bubble, so that when released from the main plant it floats to the surface. The vessel opens, forming a structure like a sail. A breeze then pushes the flower along.

This is where the female flowers come in. They also develop underwater, but the growth of spiral stalks raises them to the surface. There, they wait for male flowers to sail by. The male flower meets up with the female and pollinates it. Once fertilized, the female flower retreats below the water again and the seed-bearing fruit develops.

In case you wonder how this plant that bears no resemblance to grocery-store celery got its name, it's been said that ducks that eat a lot of it have meat that tastes like celery. Maybe that's true and maybe it isn't. Just be glad, if you have this plant in your lake, that it's there to help fish and waterfowl prosper.

BLADDERWORTS: CARNIVOROUS VEGETABLES

Does your lake contain meat-eating vegetation? It does if bladderworts are found in it. Bladderworts are among the small fraternity of carnivorous plants. Among the best-known meat-eaters are sundews and pitcher plants, which grow in the acidic soil around bogs. Bladderworts, on the other hand, can be spotted in wetlands, ponds, and in the still-water areas of some lakes.

Bladderworts are widely distributed. They tend to be small and inconspicuous; the delicate stems and leaves of some species are hard to spot even if you're right on top of them, except perhaps when they are in flower. They can grow in waters that are poor in nutrients, precisely because they can eat meat. Bladderworts come in various shapes, sizes, leaf patterns, and flower types and colors. The trait they share consists of bladders, which function as traps that capture small creatures. The bladders, sometimes hundreds or thousands of them on a single plant, grow just below the water's surface, where "hunting" is the most productive. They generally can capture one-celled critters call protozoa but also very small insects, larvae, and worms.

The bladders are ingenious feats of engineering. Each bladder consists of a sac with a "door" that only opens inward. The door is made watertight by a mucus-like substance that comes from glands on the bladder surface. Other glands inside the bladder pump water out to create something like a vacuum inside; under this condition the bladder has a concave shape as the higher water pressure outside pushes in. The door, meanwhile, is surrounded by trigger hairs that point outward, and by projections that are like insect antennae.

These antennae guide prey creatures toward the trap door. The trap is baited by a sugary substance from glands at the entrance. When a prey creature touches the trigger hairs, the tension on the door's seal is broken; water rushes in and sweeps the prey in with it. The internal glands then push the water out again, but the prey remains inside, where enzymes digest it.

There are more than two hundred bladderworts worldwide. Their names include common, great, horned, flat-leaved, large purple, small purple, northern, hidden-fruited, and creeping bladderworts. If your lake has low-pH water and is fairly nutrient-poor, then keep an eye out for these plants.

Paul Skawinski, University of Wisconsin–Stevens Point

CARING

·7·

180.5 ACRES

I like to say I live on 180.5 acres. The 0.5 is the land where the house stands. The 180 is Birch Lake. Of course I don't own the 180, but I treasure it and use it freely. It's because of the lake, the 180, that my wife and I live here. As much as we love the north, we would not have moved here if unable to live on water, and on this lake specifically. Before moving, we enjoyed the lake as tourists.

We certainly want to protect the lake, and so before building, we got familiar with the county's shoreland zoning ordinance. We did so not dreading how it would limit our use of the 0.5 acre but looking forward to complying, for the good of the 180. Some might argue that it's our 0.5 acre: we paid good money for it, so we should be able to do with it pretty much as we like; this is America. We saw it differently. I'm not about to nominate us for lake stewardship sainthood. We did what we did largely out of self-interest.

Lakes can be fragile. Development on their shores, even modest homes such as the one we built, can have long-lasting impacts. We wanted to limit ours. That includes impact on scenery—looking out and looking back. It's great to have a pretty water view looking out the lake-facing windows, but we also considered the view of our place from out on the lake. Many of us who settled on northern lakes did so for the wooded and otherwise natural shorelines and the wildlife and scenery that go with them.

An eagle perched on a tall white pine or soaring over miles of forest. A loon couple and a downy chick paddling along, the adults alternately diving for their finned food. The riot of reds, yellows, and oranges on autumn trees reflected in the water. The stark skeletons of winter hardwoods etched against white, the conifers more beautiful for simply having stayed the same. When on the lake, we wanted to see trees, not a house and lawn, even if our own. So we cleared trees only as the builders required, keeping what we could. We chose to forgo the viewing corridor we could have created by thinning a strip of woods—you can barely see the house from the lake. We could have expanded our lake vista while leaving a mostly natural shoreline. That would have taken nothing more than judicious tree thinning and trimming with help from an arborist. We simply chose not to do that.

Then there's the impact on water quality, largely from runoff that increases when homes and garages are built and driveways and walkways

paved. In light of all this, we have a fully inspected and properly maintained septic system. We use no fertilizers or pesticides. The house is
set back more than the legal minimum 75 feet from the water.

That's not to say we're perfect. Since our lot is small, we may have a
fairly high percentage of impervious surface, possibly enough to warrant
a little runoff mitigation. If that's the case, we're open to doing it. All
this is for ourselves—it doesn't even count the interests of our lakeshore
neighbors, many of whom, by the way, have made similar decisions in
how they treat their properties.

Of course, there's more to consider than our own and our lake
neighbors' sense of beauty. The lakes belong to everyone, and that includes resort patrons and day visitors. They could choose to vacation
anywhere; they come to the lakes for the natural beauty and quiet
splendor. Then there are the lakes themselves—complex and fragile
ecosystems.

In short, shoreland zoning provided valuable guidance to us as new
lake lot owners. It told us the minimum we had to do in the name of protecting the resource we care about. We could then do better than that.
And why would we not? Why would we do something on our 0.5 acres
that would harm the 180?

WATER QUALITY: WHAT'S IT WORTH?

Not long ago a new couple bought a cottage along my private road here
on Birch Lake. When I asked the husband why they chose this particular
lake, he replied, "The water is so clear." And that speaks volumes about
the importance of lake water quality. We would all rather live or vacation
on a lake that's clean and clear, with a healthy fishery, no nuisance weeds
and algae, no invasive species, and great Northwoods scenery. In fact,
people pay a premium for property on such lakes.

If you doubt this, consider two studies from Wisconsin. The Tainter
Lake study spanning 1999 to 2010 looked at 3,186 real estate transactions
on seven lakes. It found that properties on lakes with good water quality
had values two to three times higher than those on lakes with poor-
quality water. Another study on Delavan Lake, extending from 1987 to
2003, found that improved water quality after an extensive lake rehabili-
tation program led to 70 percent higher property values than on nearby
lakes that had not been restored.

Of course, to a large extent we know about this intuitively. Unless we inherited a lake home or cabin from family, we chose a lake on which to vacation and ultimately buy property in large part for the quality of the resource. And according to various studies, that quality includes all the attributes mentioned above. So it becomes pretty clear that when we do things like apply best practices to limit runoff of nutrients and sediment into the water, install buffer strips of natural vegetation along the shoreline, limit the cutting of trees to preserve natural lakeshore scenery, and do our part to help prevent the introduction of invasive species, we aren't just doing the environment a favor. We're helping to sustain the attributes we treasure, and to build and sustain a financial legacy for ourselves and our families. In fact, we're helping create a legacy for everyone who enjoys the lakes.

DARK AT NIGHT: WHAT A CONCEPT!

How dark is it on your lake after the sun goes down? Is it truly dark so that the sky blazes with stars? Or do lights on the ends of piers or lights leading up people's stairs obscure the lights in the sky? On many lakes, the latter is true. Years ago, after four years of attending college in the middle of a city, I drove to a cabin in Michigan's Upper Peninsula where my family was vacationing. The cabin had no electricity; interior light came from kerosene and gas-mantle lamps. The lake was also rather isolated and the cabins on its shore widely spaced.

On arriving, toward midnight, I pushed out onto the lake in a rowboat. My eyes immediately were drawn upward. I couldn't remember seeing stars so bright and so numerous. I fully appreciated for the first time the spectacle of the Milky Way; I easily picked out favorite constellations. I gazed at the sky for a long time in awe. Such views are becoming rare these days because we insist on lighting the places where we live, even when we don't really need the light.

It's a problem known as light pollution. It is most acute in cities, but even in the lake country, it can be very much in evidence. It takes just a few bright lights along a lake shoreline—a couple of pier lights and yard lights left on—to create a glare that impedes the view of the night sky. Turn those lights off and the stargazing could be spectacular. I wonder if it would be possible for lake groups to organize an occasional lights-out night, where all lights go off after, say, 10:00 p.m. It seems like an idea worth exploring.

TAKING CARE OF THE FISH

If you're like many lake dwellers, you're protective of your lake's fish. When angling, you may save a few for the pan, but fewer than the bag limits allow. You're careful not to do things that would harm spawning areas or otherwise hurt the fish population. Along that line, here are a couple of points to consider.

The dark side of catch and release

Catch and release is good for sustaining fish populations, but it isn't without risk. If we don't catch and release properly, then on a good day of fishing we might kill more fish than if we took the legal maximum home to filet and fry.

Yes, catch and release can have a dark side. I've considered myself a responsible angler, but in researching this topic recently I've seen that some of my habits need changing. While it's hard to know definitively how well fish survive being caught—the science is not settled—some conventional wisdom is likely wrong. For example, many of us were taught long ago that if a fish is hooked deep, such as in the throat, we should just cut the line and let it go; the hook will soon corrode and the fish will be fine. However, recent studies indicate that's not exactly so. Today's hooks are made of materials that rust or corrode slowly if at all and are likely to stay put for a long time.

If a hook's point and barb are buried in the throat, the protruding shank may block the esophagus and keep the fish from swallowing food. Some experts recommend using a strong wire cutter to snip off the exposed part of the hook. There's also a technique for going in through a gill and safely removing the hook (likely not something to try without a good hands-on demonstration from an expert).

There's also research saying that even if a fish swims off briskly, that does not mean it will survive. Sometimes deaths occur much later. Research reinforces the importance of a few good basic practices. Handle fish gently. If possible, unhook them while still in the water. Avoid using a landing net. Handle fish with wet hands. Don't lift fish by the gills. If you must have a picture taken, make it quick and cradle the fish in a horizontal position.

The right gear is also important. Use more artificials; single hooks are better than trebles. If using live bait, ignore the advice you got as a child to "let him run with it." Set the hook promptly so you hook the fish in the lip. Use barbless hooks or circle hooks. Don't fish with too light of tackle in the interest of having a long fight. That may be fun for you, but a fish "played" to exhaustion may well die later. The best course is to read up on this topic, discover the current best practices, and see where we might have room to improve our habits.

Going nontoxic

What happens to the lead sinker that slips off your fishing line, or to the worn-out lead split shot that you (heaven forbid) casually toss into the lake? Most times it probably sinks into the mud or sits on a gravel bottom, there to remain essentially forever. But "most times" isn't always. There's a chance that bit of lead will do serious damage, such as by killing a loon, a goose, a duck, or a snapping turtle. A loon, for example, may pick up the bit of lead along with small stones that help grind up food and aid digestion. An eagle or great blue heron may eat a fish that swallowed a small sinker or jig. In either event, because lead is highly toxic, the bird is likely to die.

Lead fishing tackle isn't as big an issue as lead shot used for hunting, which is known to have killed multitudes of waterfowl. Relatively few lead sinkers get lost or discarded in lakes. Still, studies document that lead tackle does kill birds, and loons in particular. How many is a matter of some debate. In a New Hampshire study published in 2000, a researcher reported that 44 percent of adult loons found dead had ingested lead tackle. On the other hand, a study in Minnesota blamed lead tackle for 5.4 percent of loon deaths. Between those extremes, a fifteen-year study in Michigan examined 186 dead loons and found that lead poisoning, mainly from jigs, caused 24 percent of the deaths.

If lead is proven to kill even a few loons, it seems anglers shouldn't hesitate to use sinkers made of nontoxic metals, such as tin or bismuth. The most obvious reason they do hesitate is that some alternatives, notably tin, don't perform as well. To get the same result as a lead sinker, it takes a bigger one made of tin. (For a similar reason, hunters for a long time resisted using steel shot instead of lead: it didn't have the same killing power.) Another point of resistance is that nontoxic sinkers and jigs cost more, especially the bismuth variety. Still another is that nontoxic tackle, though more prevalent now than years ago, is harder to find than lead, which anglers have used almost for eternity.

Momentum is growing for nontoxic tackle. Certain fishery reserves and research lakes require it. A broad-based public education campaign,

"Get the Lead Out," is active in some states; it encourages anglers to use lead alternatives. So when each fishing season approaches, we all have to choose: stick with the old ways, or make a change to help protect the wild ones.

GETTING WARMER

I've never understood why or how climate change became a partisan issue, left versus right. After all, whether climate is changing is a matter of science, of data, not political persuasion. Let's look at some data, specifically about how scientists say gradual warming is affecting and is likely to affect our lakes and fisheries.

A 2013 study showed that annual average temperatures across the Midwest increased by 1.5 degrees F between 1895 and 2012. Most of the warming seems to be occurring in winter. On Lake Mendota in Madison, Wisconsin, where ice has been watched closely for decades, the duration of ice cover has declined by eighteen days over the past century. Meanwhile, a report last fall from the U.S. Geological Survey (USGS) showed that lakes in Wisconsin have become warmer over the past thirty years as global temperatures have risen. The study, based on computer modeling, looked at the relationships between water temperature and suitability for walleye and largemouth bass in more than 2,100 Wisconsin lakes. It projects that warming will change populations of those species, partly for good and partly for ill, depending on which species you happen to prefer. Presumably, the effects would be similar for lakes in other states at similar latitude.

The USGS report observes that water temperature strongly affects walleyes and largemouths. Generally, warming water means bad news for walleyes, which prefer colder conditions, and good news for bass. The study notes that walleyes have been declining in lakes across Wisconsin for the past thirty years, while largemouth populations have increased. The researchers expect walleyes to keep declining as waters warm over time.

The news isn't all bad. The researchers pointed out that because the lakes they studied are diverse in depth, clarity, habitat, and other features, they'll respond differently to climate change; some will warm more than others. The study predicts that lakes able to sustain natural walleye reproduction will fall from 10 percent to fewer than 4 percent of Wisconsin lakes by 2050, while lakes hospitable to largemouths will increase from

60 to 89 percent. On the other hand, according to the study, walleyes in large lakes will tolerate warming better than those in smaller lakes. That means the total lake area where walleyes can reproduce will decline much less sharply. At the same time, the researchers predict that warming water will improve conditions for bass in more than five hundred lakes. That would mean more sport for bass anglers.

That's a quick look at a couple of studies on some possible impacts from a warming planet. Climate change is complex. The best any of us can do is look at the science and consider what the data tells us. And if we want to help address the issue, we can act by using less and cleaner energy: carpooling, buying a hybrid vehicle, installing LED light bulbs in our homes, subscribing to utility green energy programs—any number of actions that reduce our energy and carbon footprint.

THE BEST GIFT YOU CAN GIVE YOUR LAKE

Here on Birch Lake we're used to good but not great water clarity. It's not the best lake for snorkeling, for example, but it isn't bad. What we're not used to are persistent algae blooms. In all the years my family has come here, we have seen only a few, and they have been mild. In the heat of late August or early September, the water takes on a slight green cast and there's just a little green skin atop the still water near shore. A few days of cooler weather and it's all gone.

One recent year was different, though. There was more of the green stuff, it lasted longer (more than two weeks), and during some of that time the water had a pea-soupy look. Blue-green algae blooms, also called harmful algae blooms (HABs), are caused by photosynthetic bacteria (cyanobacteria). These organisms can build up into dense mats that emit foul odors. Blue-green algae blooms can harm a lake's ecosystem by reducing dissolved oxygen, blocking the sunlight beneficial algae need to grow, and producing toxins that can harm fish and other organisms. Those toxins can also make people and pets sick and cause allergic reactions, such as rashes.

Algae blooms have several interrelated causes. The first and most important is an excess of nutrients, most notably phosphorus. Others include extensive sunlight and long periods of warm temperatures and still water. Another possible factor is upwelling of nutrient-rich water from the lake's colder bottom layer.

If you're looking to do something on your lake to help prevent nu-trient enrichment and algae blooms, probably nothing is better than creating a natural shoreline, also called a shoreline buffer. When we build homes and cottages on lakes, we have an impact on water quality. The net effect of development is to increase runoff into the lake. That runoff can contain sediment, nutrients, and other pollutants that other-wise would not find their way into the water. If we clear the land, plant a lawn, and mow down to the water's edge, we create a clear pathway for runoff.

A shoreline buffer erects a barrier that helps keep the runoff on the land. It captures sediment. Phosphorus and nitrogen that flow with runoff into the shoreline vegetation end up promoting plant growth on land instead of contributing to weed growth and algae blooms in the lake. In addition, the buffer area provides a haven for mammals, birds, reptiles, amphibians, and insects—a little wildlife sanctuary on the property.

You might ask why ordinary grass wouldn't also stop sediment and take up nutrients. To some extent, it does, but common lawn grasses like Kentucky bluegrass and fine fescue have shallow roots, unlike the plants that grow naturally along the water's edge. Because they do not grow well in the wetter soil along shore, lawn grasses in that area tend to be thin, allowing sediment and nutrients to pass through.

A shoreline buffer is essentially a strip of native plants that you do not mow. A strip just 10 to 15 feet wide has benefits, but a width of 25 to 50 feet is much better. The buffer can help restore many critical lake-protective functions that hard structures, lawn sod, and mowing tend to remove. You can keep a buffer strip on your frontage and still have a lawn if you want one. It's worth considering the benefits of a buffer that go beyond protecting your lake's waters. For example, a shoreline buffer saves time you would otherwise spend mowing and trimming lawn, giving you more time to enjoy the lake. It gives you extra privacy and helps muffle the noise of boats and personal watercraft.

A buffer also discourages messy visitors like geese and helps keep invasive plants like purple loosestrife from getting established. It helps make your property more beautiful and attractive to home buyers and so more valuable. If you wonder how an unmowed swath of land could be beautiful, imagine that buffer strip planted with shrubs mixed in with wildflowers that bloom at different times, imparting texture, color, and variety to your lakefront landscape.

Buffer zones are important because the development of our lake shorelines continues to increase. For example, a 1965 study of northern Wisconsin lakes showed 25 percent of the shorelines developed. The state Department of Natural Resources predicts that by 2025, more than 90 percent will be developed. Natural shorelines can play a key role in mitigating the impacts of all those homes and cabins.

If you want to assess where excessive runoff is occurring on your property, consider doing it after a downpour (or while it's still raining, so long as you aren't exposed to lightning). Signs of rampant runoff include places where grass or weeds have been matted down in the direction of the water's flow, or where bare soil shows signs of erosion. These might indicate places to install some sort of diversion or an absorbent rain garden.

WHO OWNS THE WATERS?

A few years ago I was snorkeling in a small lake and was just several yards out from a private pier because I wanted to observe the fish around a crib I knew was there. The owner came down from his cottage and told me that I was trespassing—that it was illegal to come within a certain distance of a privately owned shoreline. I politely swam off even though I knew the landowner was mistaken. I had every right be where I was, doing what I was doing; I just didn't want to have an argument when there were other fish-attracting features to see on the lake.

Why was it all right for me to snorkel close to someone's pier? It's because while land fronting on lakes can be privately owned, the waters themselves belong to all the people, under a concept called the Public Trust Doctrine. It's important to know and understand this doctrine as part of the impetus to keep our waters clean and healthy.

The Public Trust Doctrine in one form or another is included in the constitutions of a number of states. It says that navigable waters are to be held in trust for the public. Departments of natural resources and similar state agencies are charged with protecting these resources. Private landowners have the right to access to the water. However, in cases where conflicts have occurred between private and public rights, courts have tended to rule on the side of the public.

So, while I had no right to touch that cottage owner's pier or to set foot on his property, I had a right to be anywhere on the lake itself, as long as I stayed in the water. The Public Trust Doctrine was at first intended to protect public rights to transportation on navigable waters, but over the years, courts have broadened it to include public rights to water quality and quantity, recreation, and scenic beauty. It has been repeatedly upheld, defined, and expanded by court cases and state laws. It is part of the culture in water-rich states.

Think about what this doctrine means. In Wisconsin, where I live, it means that the public owns the waters of Lake Michigan, the Wisconsin River, Lake Winnebago, several large flowages, and all of the state's more than 15,000 lakes and 84,000 miles of streams. We have the right to swim, boat, fish, hunt, skate, watch wildlife, and generally enjoy these waters. If we accept the Public Trust Doctrine as a right, then it comes with a responsibility we all share to use the waters wisely and help safeguard their health for the good of everyone.

A SALTY DILEMMA

Here's something to ponder as winter approaches: What happens to all the salt we apply to our roads, driveways, and sidewalks to keep them ice-free? Does that salt find its way into our waters? And if so, what effect is it having? This has become a hot topic in environmental circles. The most obvious thing that happens to salt once it's put on pavement is that it dissolves in water. Chemically, salt is sodium chloride, a molecule with one atom of sodium and one atom of chlorine. Once the snow and ice melt, that salt readily dissolves into sodium and chloride ions and is washed away.

Unfortunately, it's not really gone. In some cases, such as when applied near bridges and culverts, it can wash directly into a stream. Otherwise it goes along when water soaks into the soil. From there it eventually finds its way into groundwater. Groundwater doesn't sit still; it moves, although slowly. And since groundwater is a major source of the water in many of our lakes, the salt can show up there. Chloride is a water pollutant that in high concentrations can harm fish and water plants. Once it's in the water, there is no way to get it out.

Salt in our lakes is not yet an ecological problem comparable in severity to phosphorus pollution or invasive species, but in some lakes the chloride content is trending upward. One study in Wisconsin analyzed data from the early 1980s to 2014 and found salt levels staying essentially the same in lakes not near major highways. On the other hand, two lakes along an often-salted state highway showed chloride levels rising steadily.

One small, clear lake near a highway saw its chloride level increase from just over 2 parts per million to just under 10 parts per million over the three decades. In a much larger and very deep lake, chloride increased from about 1.4 to about 2.0 parts per million. These are still low levels; for comparison, the lakes around the state capital, Madison, have chloride levels from just under 50 to just over 100 parts per million, and trending upward. Chloride is not toxic at any of those concentrations, but it seems sensible in any case to be careful about how we use salt. Certainly road salt is the biggest contributor, and many cities are working to use less while still keeping the roads safe.

There are also things we can all do on our properties. For example, we can avoid salting when it's not really necessary. We can also hold the salt when the temperature outside is below 15 degrees F because salt doesn't work then—sand provides better traction at those times. And in general, we can just use less.

THE WOOD MAKES IT GOOD

In those old-time fishing stories about the great big bass everyone wants to catch but no one can, the bass always hangs out under a log. The wily lunker bass is a cliché, but the stories convey a useful truth: fish in our lakes relate to wood. Find logs lying on the bottom or a fallen tree leaning into the water and chances are you'll find fish.

Fisheries biologists increasingly see wood in the water as essential to improving fish habitat. They're not talking about cribs made of logs, since these mainly concentrate fish. They advocate whole trees in the water: the maze of branches provides the cover young fish need to escape from predators and grow into adults.

On lakes not disturbed by humans, trees and branches routinely fall into the water, and their numbers grow over time, so that hundreds of woody objects lie sprawled on the bottom and alongshore. Lake homeowners typically remove wood from their frontage to improve the swimming area or make way for piers. Trees that fall into the water they cut up and take away. As a result, many shorelines are nearly devoid of wood, and research suggests that this seriously hinders the development of healthy fisheries.

The technical term for logs and trees in the water is coarse woody habitat. Recent research in Wisconsin shows that the impact of wood on fish life can be profound. The study explored how the removal of wood from a lake would affect fish populations. The lake in question has two nearly equal lobes joined by a narrow channel. This made it easy to split the lake in two, remove wood from one lobe, and leave the other lobe alone. The lake was naturally rich in woody habitat with about 760 logs per mile of shoreline, about the amount of wood normally found on northern lakes where there are no dwellings.

From one lobe of the lake the researchers removed about 75 percent of the wood, reducing it to a level consistent with lakes with modest numbers of homes and cottages. Then they studied the fish life in the two lobes over four years. The yellow perch population collapsed in the lobe where wood was removed. The perch lost spawning habitat and the young that hatched were far more vulnerable to predators. By the end of the study, perch in the lake lobe were almost nonexistent. As for largemouth bass, with very few perch to feed on, they began to eat their own young and eat more land-dwelling creatures, such as frogs, that found their way into the lake. Their growth rates declined.

So it seems clear that the loss of coarse woody habitat is not good for fish. Still to be determined is to what extent adding wood can renew fisheries. In the meantime, fishery agencies now promote the addition of wood by installing "fish sticks"—clusters of downed trees anchored to the shore and partly or fully submerged.

Now, suppose that a storm just blew through and knocked down a tree along your waterfront. If it didn't break your pier, spoil your swimming area, or block your boat's access to the lake, chances are the best course is to leave it where it fell. There might even be a bonus in it: you could soon find a new prime fishing spot right in your own front yard.

IF THE BAD GUYS WERE BIGGER

A poster in my doctor's office shows a boy looking down at his hands, which are infested with all sorts of big, ugly, wormlike green, purple, and pink creatures. And the poster says, "If germs were bigger, would you wash your hands more?" It's useful to think of this when we consider properly taking care of the boats we move from one lake to another for fishing and other recreation. We all (I should hope) regularly remove any weeds from our boat and trailer when we take it out of a lake. Before launching at another lake, we check for and remove any hitchhikers that might be clinging.

It so happens, though, that some invasive species whose spread we need to control are not much more visible to the naked eye than the germs we should regularly wash off our hands. Two obvious examples are the spiny water flea and the larvae of zebra and quagga mussels — insidious pests that infest a growing number of lakes over a wide swath of North America. These species are native to Europe and Asia and got into the Great Lakes by way of ballast water discharges from ocean-going ships.

Spiny water fleas are a variety of zooplankton. They eat native daphnia, on which fish depend. They have spread across the Great Lakes and have prospered in some inland lakes. They can travel to new lakes by clinging to fish line, anchor ropes, and landing nets. Female spiny water fleas dry out and die once out of the water, but they produce eggs that can survive drying. Live specimens can travel in bait pails or boat livewells.

Zebra and quagga mussels essentially dominate Lake Michigan, filtering out the algae and zooplankton that form the base of the food chain. They've been found in some inland lakes as well. Adult mussels are big enough to spot clinging to a boat or trailer. The greater danger is that they produce microscopic larvae called veligers that, like spiny water fleas, can travel in water left in boats and containers. For these reasons, it's essential to drain our boat livewell, bilge, and motor before leaving a landing and to avoid transporting any water, such as in a bait bucket, from one lake to another. In addition, it's advisable to let boats and equipment dry for about five days before going to another lake. This should desiccate and kill any hangers-on.

Unlike handwashing, these are not matters of personal hygiene. Instead they are critical practices for protecting our lakes, and they only matter if everyone observes them. All it takes to spread an invasive species to a new lake is for one angler or boater to be careless. We should all be able to follow these steps, and if we don't, shouldn't we then forfeit our right to criticize others who may pollute our lakes and streams?

WHO LOOKS OUT FOR YOUR LAKE?

About twenty years ago, the rusty crayfish population exploded on Birch Lake, where I then vacationed with my family, and where my wife and I now live. In response, a group of residents formed the Friends of Birch Lake, a social club and a source of funds and labor to carry out a crayfish trapping program. The group had seriously dented the crayfish population before we bought our property and built our house. I owe much of my enjoyment of Birch Lake fishing to those volunteers; I am privileged now to be one of them.

Who takes care of your lake? Chances are, volunteers do—members of friends groups, lake associations, and lake districts. These people do the work that helps control and prevent invasive species, improve fish habitat, reduce nutrient inputs, install best practices to mitigate runoff, monitor and report lake water quality data, and a great deal more. These are busy people with families to raise, jobs to do, and homes to care for, and yet they give their time and energy to make their lakes better.

Lake improvement groups are active all across the country. Their members receive nothing in return for their work except the satisfaction of knowing they've done their part to improve the lakes they love, sometimes against formidable odds. In a time when government resources for lakes are scarce, volunteers are a critical line of defense for our waters.

Are you a lake volunteer? If you are, give yourself a little pat on the back and resolve to continue your good work. Does your lake have an association or friends group? If so, and you're not already a member, consider joining. If there's no such group on your lake, try starting one, even if only for social activities at first. Because who knows when your lake might face an issue to which a ready army of volunteers can respond? Meantime, we all owe gratitude to those who contribute time, energy, and, yes, money, to make our lakes healthier and more beautiful. Who looks out for the lakes? You do. I do. We do—every one of us. Because if we don't, who will?

TEACHING KIDS TO LOVE THE LAKE

One privilege of being a grandpa is the chance to influence young minds. These days I'm busily turning my young grandsons into lake lovers. The indoctrination advances each time they pay a visit with their mom (daughter Sonya) and dad (son-in-law Chad). It's all about immersing them, figuratively and literally, in lake life. If you have similar subversive intentions with your kids and grandkids, here are some tricks to consider.

Nothing beats time on the water

Take the little ones on long, slow boat rides around the lake. A pontoon boat is ideal. Stay fairly close to shore so you can point out things to see—the big pine that toppled into the water in a storm, the logs where painted turtles bask in the sun, the patches of lily pads with their immaculate white blooms. Shut down the motor and then just listen—to the wind, the waves, the calls of loons.

Take paddles in hand

Give the kids an intimate view of the water world. Seat them on cushions on the canoe floor amidships while you and your partner paddle at the bow and stern. Stay in the shallows. Cut right through beds of rushes and watershield. Ideally, go in the very early morning to improve the odds of seeing deer down at the lake to drink, or otters frolicking on shore and in the water.

Take them fishing

You don't need much—just a box of worms and a couple of cheap rods rigged with bobbers. With kids it has nothing to do with the size of fish, only with numbers. Go to a spot where panfish are abundant. As you unhook each fish, point out its features—the bright-yellow bellies of sunfish, the subtle hues of bluegills, the angelic fins of crappies. Keep the outing short; it's over when the first kid seems bored or complains.

Open their eyes

I know from experience what a thrill it is for a child to see clearly underwater. A swim mask reveals a whole new world to explore—the texture of the bottom sand and gravel, the shadowy world of weedbeds, the chance to pursue crayfish with a small net, close encounters with fish of different shapes and colors.

Watch sunsets

Head down to the pier as the day wanes with the kids' favorite snacks and drinks. Have a little picnic while the sun goes down and the clouds catch fire. Stay as darkness falls and watch for bats skimming the water, or hatching mayflies flailing on the surface until fish slurp them up in a swirl.

Stargaze

Choose a new-moon night and show the young ones the blazing wonders of the night sky. A strong flashlight beam is great for pointing out the obvious constellations, or for spotting a satellite creeping across the heavens.

Point out everything

Miss no opportunity to show the kids what you see, from the eagle or great blue heron overhead, to the water striders and whirligig beetles dancing on the surface. A quality magnifying glass is an excellent tool for letting the kids examine small critters you catch, or for seeing tiny animals like water fleas for the first time.

It's my ambition to do all these things with the boys; I've done some of them already. I'm fortunate to have the chance to let Birch Lake work its magic on them, the way other lakes years ago did on me. As a project for my retirement years, I could do a lot worse.

IF EVERYBODY DID

When I was a kid, one of my favorite books was *If Everybody Did*. It took a humorous approach to teaching children the consequences of bad behavior. On one two-page spread was a simple statement of something harmful a kid could do. The next spread showed, in a hilarious illustration, the result, with the simple caption, "This is what would happen if everybody did."

So, for example: "Spill tacks." Turn the page to find kids and adults scattered around a room with tacks sticking out from every part of their bodies. "Leave tracks." Turn the page to a living room, its walls, floor, ceiling, and furniture covered in muddy shoeprints. When it comes to lakes, there's a way to put a positive twist on "if everybody did." Just imagine what our lakes might be like if all of us would:

- Plant (or preserve) a buffer strip of trees and natural vegetation extending 30 feet or more up from our shoreline, as habitat for wildlife and protection from polluted runoff.

- Take other measures to limit runoff into the lake, like a rain garden, or rain barrels under the downspouts on our houses and garages.
- Be careful when using fertilizers, herbicides, and pesticides—apply them in only the necessary amounts, or don't use them at all.
- Pick up and properly dispose of pet waste.
- Make sure our septic systems are functioning properly; have them inspected annually and pumped out as recommended.
- Join our lake associations and contribute to their efforts to promote healthy vegetation, improve water quality, and enhance the fisheries.
- Contribute, where possible, to county, regional, and state organizations dedicated to improving our lakes and streams.
- Observe "clean boats" principles; volunteer one or two days a week at a landing to help educate boaters and prevent the spread of invasive species.
- Be ethical anglers, not always taking a limit of fish just because we can, learning best practices for unhooking and releasing fish, using circle hooks or barbless hooks, avoiding lead tackle.
- Talk to friends, neighbors, and family about lake protection practices.

None of these things are difficult. And yet, if we all did them, year after year, we would have cleaner and healthier lakes, protect the scenic values that attracted us to the lake country, and preserve the value of our properties.

BLAZES OF GLORY

Our lake property faces northwest, which means we get to see sunsets. The flip side of that is to face east and be able to catch sunrises. I don't know about others, but I've seldom watched the sun come up except when on early fishing outings, and even then I'm more likely to focus on a slip bobber or on a lure I'm steering through submerged weeds than to scan the sky.

Rising at dawn, most often, we're getting ready for the day—making breakfast, showering, choosing a wardrobe, pulling things together for work. Spectacular sunrises are likely to pass unnoticed, in the manner of the late-night aurora borealis we might later hear described by a friend who happened to be outside well after dark. Sunsets, on the other hand, are easily accessible. Even when days are at their longest, the sun goes down before a normal bedtime, and the best place to observe them is over water. It's a kind of reward for a day well spent, and for that matter even a day of slothfulness. No rule says the beauty of a sunset needs to be earned.

Sitting out on the pier bench as the wind goes still and the lake's surface becomes reflective, one can try to predict the quality of the display. If the sky is clear, there won't be much to see—just a reddish glow as the sun touches and then sinks below the tree line. But when the sun alternately hides behind and appears between clouds near the horizon, that's when to pay attention, to make a point of staying for the show.

Before the sun touches the treetops it's a yellow glow beneath and around the clouds—nothing truly special. Then the sun drops out of sight and, slowly, the yellow morphs to orange, then incandescent red. The more clouds in the sky, the farther and higher the color spreads. The tree line goes black; the water displays the trees' inverted images, and beneath that the sky's reflection, sometimes utterly still, other times undulating with gentle, slow-rolling waves, still other times diffused by fine ripples stirred by a breeze.

The colors reach a discernable peak, in much the way one can perceive the day on which the colors of autumn trees reach their crescendo. Then, slowly, the color fades until the clouds hanging in the sky take on an ashy gray, like the puffs of smoke that linger after a fireworks display. Sunsets come at different hours as the year proceeds. In the dead of winter our lake-facing windows, as seen from the sofa beside the fireplace, frame the dark shapes of oaks, maples, hemlocks, and white pines against a sea of glowing red. Early mornings, now and then, I may catch, but little note, similar brilliance in the sky to the east. It's for sunsets that I linger, watching the day go out in glorious style.

Acknowledgments

I worked on this book for more than three years, during which time I wrote earlier versions of most of the material for a column called "The Lake Where You Live" in the *Lakeland Times* (Minocqua) and the *Northwoods River News* (Rhinelander), both in northern Wisconsin. I'm indebted to publisher Gregg Walker, former news editor Jim Oxley, and current outdoor editor Beckie Gaskill for their support.

Many people helped me along the way with advice, encouragement, and education. That actually goes all the way back to Sister Verna, my eighth-grade teacher at Holy Redeemer Catholic School in Two Rivers, Wisconsin, who stoked my innate interest in the natural world. I'm also indebted to the late John Batha, biology professor at Carroll College (now Carroll University), for his instruction in limnology.

The Wisconsin Lake Leaders Institute has been a great influence. Dave Blunk nominated me for the program, put on by University of Wisconsin-Extension Lakes faculty members Kim Becken, Eric Olson, Patrick Goggin, Buzz Sorge, Carroll Schaal, and Mary Knipper. I learned a great deal from my Lake Leaders Crew II colleagues: Linda Anderson, Kirk Boehm, Dan Butkus, Anna Cisar, Wes Dawson, Scott Frank, Art Frieberg, Jim Giffin, Lisa Griffin, John Kennedy, Mary Marks, Katie Nicolas, Brenda Nordin, Lindsay Olson, Philip Peterson, LeeAnn Podruch, John Primozich, Vicki Funne Reed, John Richter, Floyd Schmidt, Tom Schroeder, Brad Steckart, Philip Sylla, Elizabeth Usborne, and Zach Wilson.

In fall 2016 I expanded my lake education by serving as Artist (writer) in Residence at the University of Wisconsin Center For Limnology's Trout Lake Research Station, at the invitation of Susan Knight and Terry Daulton. Trout Lake limnologists Noah Lottig and Carl Watras were generous with their time, as was Gregory Sass of the Wisconsin

Department of Natural Resources Northern Highland Fisheries Research Area.

Mike Engleson, executive director of Wisconsin Lakes, is instrumental in putting on the annual Wisconsin Lakes Partnership Convention, where each year I learn more about lakes and the difficult, challenging, rewarding, and necessary work of lake quality protection and improvement.

I've also gained insights and inspiration from another a group of passionate lake people in current and former members of the Oneida County Lakes and Rivers Association Board of Directors: Bob Martini, Bob Mott, Rick Foral, Rob Hagge, Connie Anderson, Bill Jaeger, Kathy Noel, Jean Roach, Norris Ross, Tom Rudolph, Tom LaDue, and Bob Thome, plus advisor Myles Alexander of the University of Wisconsin-Extension office in Oneida County.

Closer to home, I've immersed myself in lake study and lake improvement projects as treasurer and later education coordinator for Friends of Birch Lake, my home water. Current and former board associates are Mary Ann Doyle, Wynne Kayser, Randy Lepak, Mike Black, Eric Roell, Landy Koppa, Dan Wolfgram, and Rollie Schultz. My across-the-road neighbors and frequent dining-out partners Denny and Angie Thompson and Paul and Mardi Sachse, who live on Sand Lake, have been among the leading fans and promoters of the newspaper column.

Countless others whose names I don't know have helped me by posting volumes of material on lake life and lake science on the websites of state natural resource agencies, universities, and nonprofit agencies. I've drawn heavily on this material for ideas and for the information you find on these pages.

Several people generously lent their expertise to review the manuscript for technical accuracy and made helpful comments and suggestions. Special thanks to Paul Garrison, Susan Knight, John Bates, and Steve Avelallemant.

Eric Roell created the lake science illustrations for the book. Paul Skawinski was incredibly generous in sharing images from has extensive photo library; Dean Hall, Pamela Montz, Chris Hartleb, Linda Grenzer, Bob Bell, and Krista Slemmons also contributed photography.

Of course, this book would not exist except for the faith, confidence, and support of the staff at the University of Wisconsin Press, most notably Gwen Walker, Sheila McMahon, Anna Muenchrath, Terry Emmrich, Jennifer Conn, Andrea Christofferson, Sheila Leary, and

Dennis Lloyd. Thanks also to freelance editor Michelle Wing, who helped give the manuscript a nice coat of polish, as well as book designer Scott Lenz and cover designer Bruce Gore.

Finally, I owe substantial gratitude to my writers' group. It has been my pleasure to meet with them every other Wednesday these past few years at Little Creek Coffee Company in Arbor Vitae, Wisconsin. There's nothing like a little peer pressure to force a writer to produce and not procrastinate. There's also no substitute for a group of excellent writers willing to hold a person to a high standard. In getting a book published, there is often a fine line between rejection and acceptance, and these fellow writers, by forcing me to elevate my craft, surely helped place me on the right side of that line. And therefore, Sue Drum, David Foster, Andrée Graveley, Cheryl Hanson, and Elaine Hohensee— thank you.

Suggested Readings

Books

Borman, Susan, Robert Korth, and Jo Temte. *Through the Looking Glass: A Field Guide to Aquatic Plants*. Cooperative Extension of the University of Wisconsin-Extension and the Wisconsin Department of Natural Resources. Stevens Point: Wisconsin Lakes Partnership, 1997.

This book is a handy guide to keep in a lake home or cottage to help identify water plants encountered on lake adventures. It describes each plant in detail—origin and range, habitiat, value to the aquatic community, and more. Like any good field guide, it includes detailed illustrations to help in telling one species from another.

Brakken, James A. *Saving Our Lakes & Streams: 101 Practical Things You Can Do Today*. Cable, WI: Badger Valley Publishing, 2016.

In a nicely organized form, Brakken tells what lake property owners can do to keep lakes clean and healthy and lakefront life pleasant for everyone. The gist of his advice is simple: as lake stewards we need to think of others besides ourselves. The lakes belong to everyone, and we would all like to leave them to our kids and grandkids in as good or better shape than we found them. So go lightly. Be gentle. Give back.

Nelson, Darby. *For Love of Lakes*. Ann Arbor: University of Michigan Press, 2012.

Nelson combines the sensitivity of an artist with the insight of a scientist—he is a retired aquatic ecologist and professor. Among much else, he describes the geological history of lakes and the glaciers that formed them—you can almost visualize the glaciers advancing and hear the crunching of rock and the flowing of glacial meltwater. He takes readers on enlightening journeys to dozens of lakes he has known and loved.

Nichols, Wallace J. *Blue Mind: The Surprising Science That Shows How Being Near, In, On, or Under Water Can Make You Happier, Healthier, More Connected and Better at What You Do*. New York: Little, Brown, 2014.

You probably feel that looking at the water, sitting beside it, floating on it, or swimming in it helps clear your mind, eases tension, and inspires creativity as nothing else can. Nichols presents a strong scientific case that responding in these ways to water is an essential part of who and what we are. He worked with scientists, psychologists, athletes, explorers, artists, and others to consider: "What happens when our most complex organ—the brain—meets the planet's most abundant feature—water?" He cites studies showing that our affinity for water is hardwired into our being.

Skawinski, Paul M. *Aquatic Plants of the Upper Midwest: A Photographic Guide to Our Underwater Forests*. 2nd ed. Wausau, WI: Author, 2018.

This field guide includes descriptions and color pictures of aquatic plants. It's also printed on waterproof paper, so it can be taken in a boat or canoe or on a wading expedition without worry about it getting wet.

Websites and Apps

DIY Lake Science App. https://www.lawrencehallofscience.org/do_science_now/science_apps_and_activities/diy_lake_science.

This app helps families and educators investigate and learn about lakes and other freshwater ecosystems. It includes easy-to-use, hands-on activities with step-by-step instructions. The activity materials are widely available and inexpensive.

Lake Ice. lakeice.squarespace.com.

This website shares information about ice science, characteristics, and hazards, with an eye toward helping anglers, skaters, snowmobilers, ice boaters, and just plain explorers enjoy lakes in winter.

Lake Scientist. http://www.lakescientist.com.

This online journal covers lake science and research and serves as an interactive resource for scientists, students, and anyone interested in the study of lakes. It offers in-depth articles written by leading researchers in lake science, along with coverage of issues affecting lakes today.

North American Lake Management Society. https://www.nalms.org.

The mission of NALMS is to forge partnerships among citizens, scientists, and professionals to foster the management and protection of lakes and reservoirs. The website includes a wealth of information on these topics for anyone interested in lakes, regardless of scientific, professional, or educational background.

Society for Freshwater Science. https://www.freshwater-science.org.

The Society for Freshwater Science is an international organization that aims to promote understanding of rivers, streams, lakes, reservoirs, and estuaries and related habitats such as wetlands, bogs, fens, and lakeshore forests and grasslands. The website fosters exchange of information among society members, resource managers, policymakers, educators, and the public.

University of Wisconsin-Extension Lakes. https://www.uwsp.edu/cnr-ap
/UWEXLakes/Pages/default.aspx.
    While focused on Wisconsin, this website contains a wealth of infor-
mation about lakes and their care and preservation, useful to anyone interested in lake management and water quality improvement.

University of Wisconsin-Madison. Center for Limnology. https://limnology
.wisc.edu.
    The Center for Limnology's mission is to provide new knowledge and
information on aquatic ecosystems through research, education, outreach,
and public service. The website contains an abundance of scientific information including reports on the latest in limology research.

U.S. Environmental Protection Agency. Lake Shoreland Protection Resources. https://www.epa.gov/lakes/lake-shoreland-protection-resources.
    This website provides numerous resources to help lake groups and
others protect and restore fragile lake shorelands and promote better lakeshore stewardship by property owners.

Water on the Web. http://www.waterontheweb.org.
    Water on the Web helps college and high school students understand
and solve real-world environmental problems. It contains two sets of cur-
ricula, data from many lakes and rivers, extensive online primers, data
interpretation, Geographic Information System tools, and supporting
materials.

National and State Lake Associations

These associations are excellent resources for learning about lakes and lake
management and for networking with other lake lovers. Many of these groups
hold annual conventions.

Federation of Vermont Lakes and Ponds. http://vermontlakes.org/.
Illinois Lakes management Association. https://ilma-lakes.org/.
Indiana Lakes Management Society. http://www.indianalakes.org/.
Iowa Great Lakes Association. www.iagreatlakes.com.
Maine Lakes Society. http://mainelakessociety.org/.
Michigan Lake & Stream Associations. http://www.mymlsa.org/.
Minnesota Waters. https://minnesotawaters.org/.
New Hampshire Lakes Association. https://nhlakes.org/.
New York State Federation of Lake Associations. http://www.nysfola.org/.
Ohio Lake Association. http://ohiolakes.org/.
PA Lake Management Society. http://www.palakes.org/.
Wisconsin Lakes. www.wisconsinlakes.org.

Index

Ted J. Rulseh writes the newspaper column "The Lake Where You Live" and is active in lake-advocacy organizations, including the Wisconsin Citizen Lake Monitoring Network. The editor and publisher of several books on the Great Lakes region, he is the author of *On the Pond: Lake Michigan Reflections*. He lives in the lake-rich region of northeastern Wisconsin.